T0181300

Wireless Networks

Series editor

Xuemin (Sherman) Shen
University of Waterloo, Waterloo, Ontario, Canada

The purpose of Springer's new Wireless Networks book series is to establish the state of the art and set the course for future research and development in wireless communication networks. The scope of this series includes not only all aspects of wireless networks (including cellular networks, WiFi, sensor networks, and vehicular networks), but related areas such as cloud computing and big data. The series serves as a central source of references for wireless networks research and development. It aims to publish thorough and cohesive overviews on specific topics in wireless networks, as well as works that are larger in scope than survey articles and that contain more detailed background information. The series also provides coverage of advanced and timely topics worthy of monographs, contributed volumes, textbooks and handbooks.

More information about this series at http://www.springer.com/series/14180

Rong Zheng · Cunqing Hua

Sequential Learning and Decision-Making in Wireless Resource Management

 Springer

Rong Zheng
Department of Computing
McMaster University
Hamilton, ON
Canada

Cunqing Hua
School of Information Security Engineering
Shanghai Jiao Tong University
Shanghai
China

ISSN 2366-1186   ISSN 2366-1445   (electronic)
Wireless Networks
ISBN 978-3-319-84413-8   ISBN 978-3-319-50502-2   (eBook)
DOI 10.1007/978-3-319-50502-2

Printed on acid-free paper

This Springer imprint is published by Springer Nature
The registered company is Springer International Publishing AG
The registered company address is: Gewerbestrasse 11, 6330 Cham, Switzerland

Preface

Resource management has been a perpetual theme in wireless network design, deployment, and operations. In today's wireless systems, data demands continue to grow with a diverse range of applications from bandwidth-hungry multimedia streaming, delay-sensitive instant messaging, and online gaming to bulk data transfer. The ever-increasing needs for high-speed ubiquitous network access by mobile users are further aggravated by emerging machine-to-machine communication for home and industrial automation, wide-area sensing and monitoring, autonomous vehicles, etc. Delivery of these rich sets of applications is fundamentally limited by resource scarcity in wireless networks that manifests at various levels. For instance, spectrum scarcity has emerged as a primary problem when trying to launch new wireless services. Vendors and operators are increasingly looking into millimeter radio bands for 5G cellular standard though the spectrum was previously considered unsuitable for wider area applications. Interferences among wireless transceivers in close proximity continue to pose challenges to the delivery of reliable and timely services. Mobile devices are inherently power-constrained, demanding efficient communication schemes and protocols.

Many resource management solutions in wireless networks operate on the assumption that the decision makers have the complete knowledge of system states and parameters (e.g., channel states, network topology, and user density). When such information is unavailable or incomplete, probing or learning has to be conducted prior to the making of resource management decisions. As an example, in orthogonal frequency-division multiplexing (OFDM) systems, pilot signals are transmitted either along a dedicated set of subcarriers or a specific period across all subcarriers for channel estimation. This allows the adaption of subsequent transmissions to current channel conditions. Sequential learning, in contrast, is a paradigm where learning and decision-making are performed *concurrently*. The framework is applicable in a variety of scenarios where the utility of resource management resources follows single-parametrized independent and identically distributions, Markovian or unknown adversarial processes. It is a powerful tool in wireless resource management. Sequential learning and decision-making have been successfully applied to for resource management in cognitive radio networks, wireless LANs, and mesh networks.

However, there are significant barriers in the wider adoption of the framework in addressing resource management problems in the wireless community. We believe that this can be attributed to two reasons: First, the sequential learning theory originates from complex stochastic concepts posing a significant learning curve. Effective algorithms often rely on underlying assumptions such as stationary and independent stochastic processes. Identifying a suitable solution to a specific problem can be a tall order for beginners. Second, there is a disconnect between theory and practical constraints in real-world settings. For example, in practice, the timeliness of decision-making often trumps optimality. In contrast, the primary concerns of sequential learning literature are the convergence rate in a long run and the optimality of the sequence of actions.

This book is the first attempt to bridge the aforementioned gaps. Though the literature on sequential learning is abundant, a comprehensive treatment of its applications in wireless networks is lacking. In this book, we aim to lay out the theoretical foundation of the so-called multi-armed bandit (MAB) problems and put it in the context of resource management in wireless networks. Part I of this book presents the formulations, algorithms, and performance of three forms of MAB problems, namely stochastic, Markov, and adversarial. To the best of our knowledge, this is the first work that covers all the three forms of MAB problems. Part II of this book provides detailed discussions of representative applications of the sequential learning framework in wireless resource management. They serve as case studies both to illustrate how the theoretical framework and tools in Part I can be applied and also to demonstrate how existing algorithms can be extended to address practical concerns in operational wireless networks. We believe both the industry and the wireless research community can benefit from a comprehensive and timely treatment of these topics.

Hamilton, ON, Canada   Rong Zheng
Shanghai, China   Cunqing Hua
September 2016

Acknowledgements

We would like to thank our families and funding agencies for the continuing supports of our work. Some of the research work included is not possible without the help of our students, Arun Chhetri, Pallavi Arora, Thanh Le, Mohamed Hammouda, and Lingzhi Wang, and collaborators Dr. Zhu Han, Dr. Csaba Szepesvári, and Rui Ni.

Contents

Acronyms

AB-FPL	Adaptive Bandit Following the Perturbed Leader
AWGN	Additive White Gaussian Noise
CR	Cognitive radio
CSWA/CA	Carrier Sense Multiple Access with Collision Avoidance
DSA	Dynamic Spectrum Access
ERT	Experimental Regret Testing
EXP3	Exponential-weight algorithm for Exploration and Exploitation
FPL	Following the Perturbed Leader
IID	Independent and Identically Distributed
INF	Implicitly Normalized Forecaster
LBT	Listen-Before-Talk
MAB	Multi-armed Bandit
MDP	Markov Decision Process
POMDP	Partially Observable Markov Decision Process
SNR	Signal-to-Noise ratio
TS	Thompson Sampling
UCB	Upper Confidence Bound
WLAN	Wireless Local Area Networks

Part I
Theory

Chapter 1
Introduction

Abstract Sequential learning provides a rigorous framework to address the trade-offs between exploration and exploitation in face of uncertainty, vividly captured in the multi-armed bandit problem. In this chapter, we provide an overview of the sequential learning framework and a taxonomy of the types of problems where the framework applies.

1.1 The Gambler's Dilemma

According to urban legends, the development of probability theory originated from the ancient pastime of gambling. Thus, it is only fitting to start the discussion of sequential learning and decision-making with the story of a gambler and slot machines.

A gambler arrives in a casino with a bag of money. Among a row of slot machines, she needs to decide how many times to play each machine and in which order to play them. When played, each machine provides a random reward. The goal of the gambler is to maximize the total rewards earned through a sequence of lever pulls. This is the famed problem of *multi-armed bandit* (MAB), first considered by Robbins in 1952 [Rob52]. Clearly, if the gambler has the prior knowledge regarding which slot machine is the most "profitable" (or equivalently, with the highest reward) and the rewards from multiple pulls of the same machine are independent, she would simply pull the level from the most profitable machine repeatedly. In absence of such knowledge, the gambler can still observe the outcome of each lever pull and estimate the profitability of the machines. In this case, the gambler faces a dilemma between "exploiting" the machine that has the highest expected reward and "exploring" different machines to gain more information about their profitability.

The trade-offs between "exploitation" and "exploration" are in fact common place in real life in the face of uncertainty and/or incomplete information. In forming a long-term relationship, rarely a person dates only one eligible candidate and commits to the relationship. The opposite extreme of dating as many people as possible without ever committing is unlikely to result in lifelong happiness (with some exceptions). Or, consider a new student who just starts her PhD study and is not sure which

© Springer International Publishing AG 2016

R. Zheng and C. Hua, *Sequential Learning and Decision-Making in Wireless Resource Management*, Wireless Networks, DOI 10.1007/978-3-319-50502-2_1

research topic to work on. She may spend much time reading up literature on each potential topic well into her third year of study or quickly zone in on one that appears promising and pay little attention to the development on any other topic. As many successful PhDs would attest, either approach is a good one. Instead, one should have a good mix of depth and breath. Similarly, when investing stocks, we have been taught not to put all eggs in one basket (i.e., exploitation). On the other hand, over-diversification of one's portfolio (i.e., exploration) often leads to low yields or losses. In the context of resource management in wireless networks, consider a RF transmitter to select the best channel or carrier to operate on. The transmitter needs to decide how often it explores (or, probes) channels that it has less knowledge and how often it exploits (or, utilizes) the best known channel so far.

Sequential learning provides a rigorous framework to address the trade-offs between exploration and exploitation. It targets problems where learning and decision-making are interleaved, namely, the knowledge accrued guides future decisions, and the decisions made would affect what knowledge shall be gained. In the subsequent discussions, unless specified otherwise, we use the terms "actions", "arms", and "machines", "players" and "users", "reward" and "payoff", "policy", "strategy", and "algorithm" interchangeably.

A sequential learning strategy consists of a series of actions given the observed reward processes in time. In designing sequential learning strategies, the first and foremost question is how to measure optimality. Typically, an oracle policy that maximizes the total reward given the complete information of the reward processes is used as a baseline. Depending on the nature of the problem under consideration, such a policy may be a single-action policy, namely, there exists a single optimal action, or, a policy that alternates among multiple actions. Note that the computation complexity of finding such an oracle policy itself may sometimes be NP-hard. In evaluating the performance of a specific policy, due to the uncertainty and the stochastic nature of the reward processes, the notion of *expected cumulative regret* is often considered. Cumulative regrets are the differences in the cumulative rewards attained by running the target policy and the oracle policy. Cumulative regrets are a function of time. They can be defined in terms of discounted regrets over infinite time or simply as a sum over a finite horizon. In generally, we are interested in devising policies that achieve sublinear regrets in time. The faster the target policy converges to an optimal policy, the slower the cumulative regret grows. An *anytime optimal policy* minimizes the cumulative regret for any time horizon t. A *fixed-horizon policy* aims to minimize the cumulative regret for a given time horizon T. Finally, an *asymptotically optimal policy* is optimal as the time horizon goes to infinity.

1.2 A Taxonomy of Multi-armed Bandit Problems

There are three fundamental formulations of the bandit problem depending on the assumed nature of the reward process: stochastic, adversarial, and Markovian (Fig. 1.1). In discussing policies for MAB problems, it is common to compare them

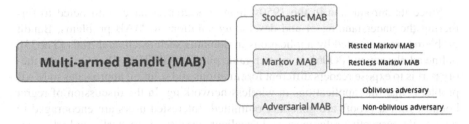

Fig. 1.1 A Taxonomy of Different MAB problems

to an optimal solution where the complete knowledge of the underlying reward process is known. Let us call this solution *the oracle optimal policy*.

In *stochastic MAB*, the reward process of each arm is independent and identically distributed (IID). It is often but not always assumed that the underlying distributions have a finite support. The rewards across different arms may be dependent. The oracle optimal policy for any stochastic MAB problem is a single-action policy that plays the arm with the highest expected reward the entire time.

In *Markov MAB*, the reward process of each arm is modeled by a Markov chain with unknown transition probabilities and an unknown expected reward in each state. Two types of Markov MAB have been investigated, namely, rested Markov MAB and restless Markov MAB. In rested Markov MAB, also called *sleeping bandit*, an arm changes its state only when it is pulled. In contrast, in restless Markov MAB, arms change states regardless of the player's action. Rested Markov MAB can be viewed as a special case of restless Markov MAB, where the state transitions of an arm follow two different Markov chains, one for the rested case and one when in play. The oracle optimal policy for Markov MAB may alternate among multiple arms and can be determined using solutions to the corresponding Markov Decision Process (MDP).

In *adversarial MAB problems*, no stochastic assumption is made on the generation of rewards. One can view the rewards as being picked by an adversary. If the adversary generates the reward process ahead of the actual plays, such an adversary is called *oblivious adversary* as it is oblivious of the player's strategy. On the other hand, at each time step, simultaneously with the player's choice of the arm, a *nonoblivious* adversary assigns to each arm a reward. In the latter case, one can view the adversarial MAB problem as a game between the player and the adversary, and thus game theoretical results such as minmax regrets are often considered.

Beside the aforementioned classification, many variants of the basic formulations exist. For example, in multiplay MAB problems [AVW87a, AVW87b, KHN15], a player can pull a limited number of arms simultaneously. In pure exploration MAB problems [BMS11, Che+14], a player aims to gain the most knowledge over t time so as to maximize the likelihood of picking the best action at time $t + 1$. In dueling bandit problems [Yue+12], the player has to choose a pair or arms in each round and can only observe the relative performances of these two arms, i.e., the player only knows which arm has the highest reward.

Since its introduction in the 1950s, many researchers have contributed to furthering the understanding of and developing solutions to MAB problems. Bandit problems are motivated by applications in domains such as clinical trials [Tho33], online ad placement [PO06], computer game playing [GWT06]. Our aim in the theory part is to expose readers different formulations and solution approaches to MAB problems that find applications in wireless networking. In the discussion of regret bounds, formal proofs are generally ommitted. Interested users are encouraged to refer to the respective references and excellent surveys on stochastic and adversarial MAB problems [BC12], and Markovia MAB problems [MT08, GGW11] for detailed proof techniques. The book by Cesa-Bianchi and Lugosi [CL06] provides a comprehensive treatment of prediction using expert advices.

1.3 Organization

In this part, we present the formulation, strategies, and regret bounds for the three forms of bandit problems. In Chap. 2, we discuss stochastic MAB and its variants. In Chap. 3, Markov MAB is discussed. Strategies for the rested and restless MAB problems are presented. In Chap. 4, we give the formulation and algorithms for adversarial MAB problems.

References

[AVW87a] V. Anantharam, P. Varaiya, and J. Walrand. "Asymptotically efficient allocation rules for the multiarmed bandit problem with multiple plays-Part I: I.I.D. rewards". In: Automatic Control, IEEE Transactions on 32.11 (Nov. 1987), pp. 968–976. ISSN: 0018-9286.

[AVW87b] Venkatachalam Anantharam, Pravin Varaiya, and JeanWalrand. "Asymptotically efficient allocation rules for the multiarmed bandit problem with multiple plays-Part II: Markovian rewards". In: IEEE Transactions on Automatic Control 32.11 (1987), pp. 977–982.

[BC12] Sébastien Bubeck and Nicol'o Cesa-Bianchi. "Regret Analysis of Stochastic and Nonstochastic Multi-armed Bandit Problems". In: Foundations and Trends in Machine Learning 5.1 (2012), pp. 1–122.

[BMS11] Sébastien Bubeck, Rémi Munos, and Gilles Stoltz. "Pure exploration in finitely-armed and continuous-armed bandits". In: Theor. Comput. Sci. 412.19 (2011), pp. 1832–1852.

[Che+14] Shouyuan Chen et al. "Combinatorial Pure Exploration of Multi-Armed Bandits". In: Annual Conference on Neural Information Processing Systems 2014. Montreal, Quebec, Canada, Dec. 2014, pp. 379–387.

[CL06] Nicolo Cesa-Bianchi and Gabor Lugosi. Prediction, Learning, and Games. Cambridge University Press, 2006.

[GGW11] John Gittins, Kevin Glazebrook, and Richard Weber. Multi-armed bandit allocation indices. John Wiley & Sons, 2011.

[GWT06] Sylvain Gelly, YizaoWang, and Olivier Teytaud. Modification of UCT with patterns in Monte-Carlo Go. Tech. rep. INRIA, 2006.

[KHN15]   Junpei Komiyama, Junya Honda, and Hiroshi Nakagawa. "Optimal Regret Analysis of Thompson Sampling in Stochastic Multi-armed Bandit Problem with Multiple Plays". In: Proceedings of the 32nd International Conference on Machine Learning, ICML 2015, Lille, France, 6–11 July 2015. 2015, pp. 1152–1161.

[MT08]   Aditya Mahajan and Demosthenis Teneketzis. "Multi-armed bandit problems". In: Foundations and Applications of Sensor Management. Springer, 2008, pp. 121–151.

[PO06]   Sandeep Pandey and Christopher Olston. "Handling advertisements of unknown quality in search advertising". In: Advances in neural information processing systems. 2006, pp. 1065–1072.

[Rob52]   Herbert Robbins. "Some Aspects of the Sequential Design of Experiments". In: Bulletin of the American Mathematical Society 58 (1952), pp. 527–535.

[Tho33]   William R Thompson. "On the likelihood that one unknown probability exceeds another in view of the evidence of two samples". In: Biometrika 25.3/4 (1933), pp. 285–294.

[Yue+12] Yisong Yue et al. "The k-armed dueling bandits problem". In: Journal of Computer and System Sciences 78.5 (2012), pp. 1538–1556.

Chapter 2
Stochastic Multi-armed Bandit

Abstract In this chapter, we present the formulation, theoretical bound, and algorithms for the stochastic MAB problem. Several important variants of stochastic MAB and their algorithms are also discussed including multiplay MAB, MAB with switching costs, and pure exploration MAB.

2.1 Problem Formulation

In stochastic MAB problems, the reward processes do not depend on the players' actions. In this case, the reward processes can be viewed as being generated by environments. A K-armed bandit is defined by random variables $X_{i,n}$ for $1 \le i \le K$, and $n \ge 1$, where each i is the index of a gambling machine (i.e., the "arm" of a bandit) and n is the number of plays of the respective machine. Successive plays of machine i yield rewards $X_{i,1}, X_{i,2}, \ldots$, which are IID according to an unknown distribution with unknown expectation μ. In its most basic formulation, the rewards are also independent across different machines, namely, $X_{i,s}$ and $X_{j,t}$ are independent for each $1 \le i < j \le K$ and $s, t \ge 1$.

A policy, or an allocation strategy, π is an algorithm that chooses the next machine to play based on the sequence of past plays and observed rewards. Let $T_i^\pi(n)$ be the number of times that machine i has been played by π during the first n plays. Denote by I_t^π the machine played at time t. To quantify the performance of a sequence of plays, the concept of *regret* is introduced. It measures the difference between the reward which could have been accumulated if the optimal action is selected, and the reward occurred by playing the actual actions according to the allocation strategy up to time n. Formally,

Definition 2.1 (*Regret*) The cumulative *regret* of policy π up to time n is defined as the difference between the rewards achieved by following the optimal action instead of choosing the actual actions $I_1^\pi, \cdots, I_n^\pi$, that is,

$$R_n^\pi = \max_{i=1,\cdots,K} \sum_{t=1}^{n} X_{i,t} - \sum_{t=1}^{n} X_{I_t^\pi,t}. \tag{2.1}$$

R. Zheng and C. Hua, *Sequential Learning and Decision-Making in Wireless Resource Management*, Wireless Networks, DOI 10.1007/978-3-319-50502-2_2

Since both the rewards and player's actions are stochastic, the following two forms of average regrets have been defined:

Definition 2.2 (*Expected regret*) The expected regret of playing I_1^π, I_2^π, ... is

$$\mathbb{E}\left[R_n^\pi\right] = \mathbb{E}\left[\max_{i=1,\cdots,K}\sum_{t=1}^{n}X_{i,t} - \sum_{t=1}^{n}X_{I_t^\pi,t}\right], \tag{2.2}$$

and

Definition 2.3 (*Pseudo-regret*) The pseudo-regret of playing I_1^π, I_2^π, ... is

$$\overline{R_n^\pi} = \max_{i=1,\ldots,K}\mathbb{E}\left[\sum_{t=1}^{n}X_{i,t} - \sum_{j=1}^{K}X_{I_t^\pi,t}\right]. \tag{2.3}$$

In both definitions, the expectation $\mathbb{E}[\cdot]$ is taken with respect to the random draw of both rewards and the actions due to the policy. Clearly, pseudo-regret is a weaker form of regret as one competes against the action which is optimal only in expectation. The expected regret, in contrast, is with respect to the action which is optimal for the sequence of reward realizations. Therefore, we have $\overline{R_n^\pi} \leq \mathbb{E}[R_n^\pi]$. In subsequent discussion, unless noted otherwise, we consider pseudo-regrets only and thus drop the term "pseudo".

Let $\mu^* \stackrel{def}{=} \max_{1 \leq i \leq K}\mu_i$, namely, the expected reward of the best arm, and $i^* \in \arg\max_{i=1,\ldots,K}\mu_i$. Equation (2.3) can be further simplified as,

$$\overline{R_n^\pi} = n\mu^* - \sum_{j=1}^{K}\mu_j\mathbb{E}\left[T_j^\pi(n)\right] \tag{2.4}$$

$$= n\mu^* - \mathbb{E}\left[\sum_{t=1}^{n}\mu_{I_t^\pi}\right]. \tag{2.5}$$

A couple of comments are in order to better understand the physical meaning of the regret defined above and the nature of MAB strategies.

Remark 2.1 From (2.5), it is easy to see that $\overline{R_n^\pi}$ is positive and nondecreasing with n. Thus, in designing MAB policies, one can at best strive for reducing the growth rate of the regret. In general, sublinear regrets (with respective to time n) are desirable as linear regrets can be trivially achieved. To see why this is the case, let us consider two naive policies. The first policy, called the *super-conservative-exploitation (SCE) policy*, simply sticks to the first machine, say I, in sight and repeatedly plays it. The regret of the SCE policy can be easily computed as $n(\mu^* - \mu_I) = \Theta(n)$. The other extreme would be a policy, called *hyperexploration-with-commitment-phobia (HECP)*, that uniformly at random chooses a machine out of the K machines at each play. Its regret can be computed as $n(\mu^* - \bar{\mu}) = \Theta(n)$, where $\bar{\mu} = \frac{1}{K}\sum_{j=1}^{K}\mu_j$ is the average expected reward. Behaviors of the two extreme policies reveal the necessity of carefully balancing the trade-off between exploitation and exploration.

Remark 2.2 A relevant question is whether sublinear regrets are generally attainable. From (2.5), we see that to achieve sublinear regrets, $T_{i*}^{\pi}(n)$, i.e., the number of times the optimal arm is played by time n should grow superlinearly with n (note that this is not true with HECP). In other words, the policy should choose the optimal arm more and more often. It is equivalent to say that the strategy converges to the optimal strategy.[1] This, intuitively, is not hard to do as long as we observe the reward from each arm *often enough* such that the best arm can be *eventually* identified and exploited. Clearly, the terms *often enough* and *eventually* need to be rigorously quantified. Nevertheless, the key insight is that there is a tight connection between stochastic MAB and parameter estimation.

2.2 Theoretical Lower Bound

From the discussion in Sect. 2.1, we aim to develop policies for stochastic MAB problems that achieve sublinear regrets. It is important to first establish the lower bound for the regrets of arbitrary policies.

In their seminar work [LR85], Lai and Robbins found, for specific families of reward distributions (Bernoulli indiced by a single real parameter), the lower bound for regret is logarithm in time. Specifically, for $p, q \in [0, 1]$, we denote by $D(p\|q)$ the Kullback–Leibler (KL) divergence between a Bernoulli of parameter p and a Bernoulli of parameter q, defined as

$$D(p\|q) = p \ln \frac{p}{q} + (1 - p) \ln \frac{1 - p}{1 - q}.$$

Let $\Delta_i = \mu^* - \mu_i, i = 1, 2, \ldots, K$ be the suboptimality parameters. The following result was proven in [LR85].

Theorem 2.1 (Distribution-dependent lower bound [LR85]) *Consider a strategy that satisfies* $\mathbb{E}[T_i(n)] = o(n^\alpha)$ *for any set of Bernoulli reward distributions, any arm i with $\Delta_i > 0$, and any $\alpha > 0$. Then, for any set of Bernoulli reward, we have*

$$\lim \inf_{n \to +\infty} \frac{\overline{R_n}}{\ln n} \geq \sum_{i : \Delta_i > 0} \frac{\Delta_i}{D(\mu_i \| \mu^*)}.$$

---

[1]This type of strategies are also called *no-regret policies*. But this term is confusing and is thus omitted here.

Theorem 2.1 is quite a remarkable result. It states any policy that plays any inferior arm subpolynomial number of times over n would incur at least logarithmic regrets asymptotically. Additionally, the constant multiplier of the logarithmic regret bound is determined by the suboptimality parameters (e.g., Δ_i) and the KL divergences between the inferior arms and the optimal one.

Let $c = \sum_{j:\Delta_i>0} \frac{\Delta_i}{D(\mu_i\|\mu^*)}$. From Pinsker's inequality and the fact that $\ln x \leq x - 1$, it is easy to show that

$$2(\mu_i - \mu^*)^2 \leq D(\mu\|\mu^*) \leq \frac{(\mu_i - \mu^*)^2}{\mu^*(1-\mu^*)}.$$

Therefore,

$$\sum_{j:\Delta_i>0} \frac{\mu^*(1-\mu^*)}{\Delta_i} \leq \sum_{j:\Delta_i>0} \frac{\Delta_i}{D(\mu_i\|\mu^*)} \leq \sum_{j:\Delta_i>0} \frac{1}{2\Delta_i} \tag{2.6}$$

Consider that the arms are arranged in the descending order of their expected rewards, namely, $\mu^* = \mu_1 \geq \mu_2 \geq \ldots \mu_K$, and, $0 = \Delta_1 \ll \Delta_2 \ll \ldots \ll \Delta_K$. From (2.6), we can see that c is roughly bounded from below by $\frac{\mu^*(1-\mu^*)}{\Delta_2}$, where Δ_2 is the difference between the expected rewards of the best and the second best arm. This should come at no surprise because if the expected rewards of the top two arms are similar, it is difficult to distinguish them through few observations. Thus, the lower bound reveals a key property of sequential learning, namely, the smaller the difference between the expected rewards of the top two actions, the slower it takes for a policy to converge.

One may question that the subpolynomial condition on the number of times inferior arms played in Theorem 2.1 is too restrictive. The answer is negative. As will be discussed in subsequent sections, there exist policies that achieve the same order in regrets as the lower bound (also called *order-optimal policies*). Clearly, these policies satisfy the subpolynomial condition as well. Furthermore, the condition in fact provides insights on the design of good policies—the number of times that inferior arms played should be kept subpolynomial.

2.3   Algorithms

One principle in designing sequential learning policies for the stochastic MAB problem is the so-called *optimism in face of uncertainty*. Consider the player has accumulated some data on the environment and must decide which arm to select next. First, a set of "plausible" environments (or hypothesis) that "agree" with the data are constructed. Next, the most "favorable" environment is identified in the set. Then, the decision that is the most optimal in this most favorable and plausible environment shall be made. We will make concrete the different aspects of the principle through the discussion of policies for stochastic MAB.

2.3.1 Upper Confidence Bound (UCB) Strategies

Our discussion of UCB strategies follows that of [BC12], which assumes that the distribution of rewards X satisfies the following moment conditions. There exists a convex function ψ on the reals such that, for all $\lambda > 0$,

$$\ln \mathbb{E}\left[e^{\lambda(X-\mathbb{E}[X])}\right] \leq \psi(\lambda) \text{ and } \ln \mathbb{E}\left[e^{\lambda(\mathbb{E}[X]-X)}\right] \leq \psi(\lambda). \tag{2.7}$$

When the reward distributions have support in $[0, 1]$, one can take $\psi(\lambda) = \frac{\lambda^2}{8}$. In this case, (2.7) is also known as Hoeffding's lemma. Denote $\psi^*(\varepsilon)$ the Legendre–Fencher transform of ψ, namely,

$$\psi^*(\varepsilon) = \sup_{\lambda \in \mathbb{R}}(\lambda \varepsilon - \psi(\lambda)).$$

Let $\hat{\mu}_{j,s}$ be the sample mean of rewards obtained by pulling arm j for s times. In distribution, $\hat{\mu}_{j,s}$ is equal to $\frac{1}{s}\sum_{t=1}^{s} X_{j,t}$. The (α, ψ)-UCB strategy, where $\alpha > 0$ is an input parameter, is an index-based policy. The index is the sum of two terms in (2.8). The first term is the sample mean of rewards obtained so far. The second term is related to the size of the one-sided confidence interval for the sample mean within which the true expected reward falls with high probability.

Algorithm 2.1: (α, ψ)-UCB

1 **Init:** *Play each machine once.*
2 **Init:** $T_i(K) = 1, j = 1, 2, \dots K.$
3 **for** $t=K+1, K+2, \dots, n$ **do**
4

$$I_t = \arg\max_{i=1,\dots,K}\left[\hat{\mu}_{j,T_i(t-1)} + (\psi^*)^{-1}\left(\frac{\alpha \ln t}{T_i(t-1)}\right)\right]. \tag{2.8}$$

5   Play arm I_t and observe $X_{I_t,t}$.

From Markov's inequality and (2.7), one obtains that

$$\mathbb{P}\left(\mu_j - \hat{\mu}_{j,s} > \varepsilon\right) \leq e^{-s\psi^*(\varepsilon)}, \forall j \tag{2.9}$$

In other words, with probability at least $1 - \delta$,

$$\hat{\mu}_{j,s} + (\psi^*)^{-1}\left(\frac{1}{s}\ln\frac{1}{\delta}\right) > \mu_j. \tag{2.10}$$

Set $\delta = t^{-\alpha}$ and $s = T_j(t-1)$ in (2.10). With probability at least $1 - t^{-\alpha}$, the following holds:

$$\hat{\mu}_{j,T_j(t-1)} + (\psi^*)^{-1}\left(\frac{\alpha \ln t}{T_j(t-1)}\right) > \mu_j. \tag{2.11}$$

Furthermore, (2.10) holds with probability at least $1 - t^{-\alpha}$, which diminishes as t grows for $\alpha > 0$.

Revisiting the optimism in face of uncertainty principle, we see that the (α, ψ)-UCB policy constructs based on past observations, hypotheses of the expected reward of each action ("plausible environments"), and picks the action with the highest plausible rewards ("the most optimal in the most plausible environments").

It has been proven in [BC12] that the (α, ψ)-UCB policy achieves logarithmic regrets. Since the proof technique is representative of UCB-like policies, we provide a proof sketch here.

Theorem 2.2 (Pseudo-regret of (α, ψ)-UCB [BC12]) *Assume that the reward distributions satisfy* (2.7). *Then* (α, ψ)-*UCB with* $\alpha > 2$ *satisfies,*

$$\overline{R_n} \leq \sum_{i:\Delta_i > 0} \left(\frac{\alpha \Delta_i}{\psi^*(\Delta_i/2)} \ln n + \frac{\alpha}{\alpha - 2} \right). \tag{2.12}$$

Proof The main idea is to show that suboptimal arms are played at most logarithmic number of times among n plays. If $I_t = i$, one of the following three inequalities must be true for any suboptimal arm i,

$$\hat{\mu}_{i^*, T_{i^*}(t-1)} + (\psi^*)^{-1}\left(\frac{\alpha \ln t}{T_{i^*}(t-1)} \right) \leq \mu^* \tag{2.13}$$

$$\hat{\mu}_{i, T_i(t-1)} > \mu_i + (\psi^*)^{-1}\left(\frac{\alpha \ln t}{T_i(t-1)} \right) \tag{2.14}$$

$$T_i(t-1) < \frac{\alpha \ln t}{\psi^*(\Delta_i/2)}. \tag{2.15}$$

To see why this is true, assume all inequalities are false. We have

$$
\begin{aligned}
\hat{\mu}_{i^*, T_{i^*}(t-1)} + (\psi^*)^{-1}\left(\frac{\alpha \ln t}{T_{i^*}(t-1)} \right) \quad &> \quad \mu^* \\
&= \quad \mu_i + \Delta_i \\
\overset{\text{negation of (2.15)}}{\geq} \quad &\mu_i + 2(\psi^*)^{-1}\left(\frac{\alpha \ln t}{T_i(t-1)} \right) \\
\overset{\text{negation of (2.14)}}{\geq} \quad &\hat{\mu}_{i, T_i(t-1)} + (\psi^*)^{-1}\left(\frac{\alpha \ln t}{T_i(t-1)} \right)
\end{aligned}
$$

The last inequality implies $I_t \neq i$ and thus a contradiction.

Let $l = \left\lceil \frac{\alpha \ln n}{\psi^*(\Delta_i/2)} \right\rceil$. It is easy to show that

$$T_i(n) = \sum_{t=1}^{n} \mathbb{I}_{\{I_t=i\}}$$

$$\leq l + \sum_{t=1}^{n} \mathbb{I}_{\{I_t=i, T_i(t-1) \geq l\}}$$

$$\leq l + \sum_{t=l+1}^{n} \mathbb{I}_{\{(2.13)\}} + \mathbb{I}_{\{(2.14)\}}$$

Taking expectation on both sides, we have

$$
\begin{aligned}
\mathbb{E}\left[T_i(n)\right] \leq{}& l + \sum_{t=l+1}^{n} \mathbb{P}\left(\hat{\mu}_{i^*,T_{i^*}(t-1)} + (\psi^*)^{-1}\left(\frac{\alpha \ln t}{T_{i^*}(t-1)}\right) \leq \mu^*\right) \\
&+ \sum_{t=l+1}^{n} \mathbb{P}\left(\hat{\mu}_{i,T_i(t-1)} > \mu_i + (\psi^*)^{-1}\left(\frac{\alpha \ln t}{T_i(t-1)}\right)\right) \\
\overset{\text{by union bound}}{\leq}{}& l + \sum_{t=l+1}^{n} \sum_{s=1}^{t} \mathbb{P}\left(\hat{\mu}_{i^*,s} + (\psi^*)^{-1}\left(\frac{\alpha \ln t}{s}\right) \leq \mu^*\right) \\
&+ \sum_{t=l+1}^{n} \sum_{s=1}^{t} \mathbb{P}\left(\hat{\mu}_{i,T_i(t-1)} > \mu_i + (\psi^*)^{-1}\left(\frac{\alpha \ln t}{T_i(t-1)}\right)\right) \\
\overset{\text{by (2.11)}}{\leq}{}& l + 2 \sum_{t=l+1}^{n} \sum_{s=1}^{t} t^{-\alpha} \\
\leq{}& l + 2 \sum_{t=l+1}^{\infty} t^{-\alpha+1} \\
\leq{}& \frac{\alpha \ln n}{\psi^*(\Delta_i/2)} + \frac{\alpha}{\Delta_i(\alpha-2)}.
\end{aligned}
$$

Finally, since $\overline{R_n} = \sum_{j:\Delta_i>0} \Delta_j T_i(n)$, we have the statement in the theorem. □

When the rewards are characterized by $[0, 1]$-valued random variables, let $\psi(\lambda) = \frac{\lambda^2}{8}$ and thus $\psi^*(\varepsilon) = 2\varepsilon^2$. Substituting the corresponding term in (2.12), we obtain the following regret bound for the UCB policy:

$$\overline{R_n} \leq \sum_{j:\Delta_i>0} \left(\frac{2\alpha}{\Delta_i} \ln n + \frac{\alpha}{\alpha-2}\right). \tag{2.16}$$

This gives rise to the α-UCB policy, which at time t plays the arm

$$I_t = \arg\max_{j=1,\dots,K} \hat{\mu}_{j,T_i(t-1)} + \sqrt{\frac{\alpha \ln t}{2T_i(t-1)}}.$$

It is identical to the UCB1 policy first proposed by Auer et al. with $\alpha = 4$ in [ABF02].

Algorithm 2.2: UCB1-normal Policy

1 **for** $t = 1, \ldots, n$ **do**
2   Play any machine that has been played less than $\lceil 8 \log t \rceil$ times.
3   Otherwise, play machine i that maximizes

$$\hat{\mu}_i + \sqrt{16 V_{i,T_i(t-1)} \cdot \frac{\ln(t-1)}{T_i(t-1) - 1}}. \tag{2.17}$$

     $T_i(t) = T_i(t-1) + 1$;
4   Update $\hat{\mu}_i$ and $V_{i,T_i(t)}$ upon observing the reward.

Another interesting special case is when the rewards follow Gaussian distributions with standard deviations bounded above by σ_{max}. In this case, using the moment generation function, one can show that $\psi(\lambda) = \frac{1}{2}\sigma_{max}^2 \lambda^2$ and thus $\psi^*(\varepsilon) = \frac{1}{2}\frac{\varepsilon^2}{\sigma_{max}^2}$. This allows us to derive a α-normal-UCB policy with a logarithmic regret bound. However, the inclusion of the maximum standard deviation term σ_{max} requires prior knowledge of the reward processes, which may not always be available. This difficulty can be attributed to the fact that the (α, ψ)-UCB policy only takes into account sampling means. Consideration of sampling variances gives rise to a class of UCB policies with tighter regret bounds [AMS09]. In what follows, we only discuss one such policy called UCB-normal proposed by Auer [ABF02].

Consider the rewards X of each action follow normal distribution with unknown mean and variance. The UCB1-normal policy (Algorithm 2.2) utilizes the estimation of sample variances as $V_{j,s} = \frac{1}{s}\sum_{t=1}^{s}(X_{j,t} - \hat{\mu}_{j,s})^2 = \frac{1}{s}\sum_{t=1}^{s} X_{j,t}^2 - \hat{\mu}_{j,s}^2$. Theorem 2.3 states the regret bound of the UCB1-normal policy. The correctness of the results is based on certain bounds on the tails of the χ^2 and the Student's t-distribution that can only be verified numerically.

Theorem 2.3 (Regret bound for UCB1-normal [ABF02]) *For all $K > 1$, if policy UCB1-normal is run on K machines having normal reward distributions P_1, $P_2, \ldots, P_K$, then its expected regret after any number n of plays is at most*

$$256 \left(\sum_{i:\Delta_i > 0} \frac{\sigma_i^2}{\Delta_i} \right) \log n + \left(1 + \frac{\pi^2}{2} + 8 \log n \right) \sum_{j=1}^{K} \Delta_j,$$

where $\sigma_1^2, \sigma_2^2, \ldots, \sigma_K^2$ are the variances of the normal distributions $P_1, P_2, \ldots, P_K$.

2.3.2 ε-Greedy Policy

A simple and well-known heuristic for the bandit problem is the so-called ε-greedy rule. The idea is to play with probability $1 - \varepsilon$ the machine with the highest average reward, and with probability ε a randomly chosen machine. ε is also called the *exploration* probability. Clearly, with a fixed ε, the average regret incurred will grow linearly. A simple fix is to let ε go to zero at a certain rate. In [ABF02], Auer show that a rate of $1/t$, where t is the index of the current play, allows for a logarithmic bound on the regret. The resulting policy, ε_n-greedy is described in Algorithm 2.3.

Algorithm 2.3: ε_t-Greedy Policy

Input: $c > 0$ and $0 < d < 1$

1 **Init**: *Define the sequence* $\varepsilon_t \in (0, 1]$, $t = 1, 2, \ldots$, *by*

$$\varepsilon_t \stackrel{def}{=} \min\left\{1, \frac{cK}{d^2t}\right\}.$$

   for *t = 1, 2, ..., n* **do**

2 | $I_t = \arg\max_{j=1,\ldots,K} \hat{\mu}_j$.

3 | Draw u uniformly from $[0, 1]$.

4 | **if** $u > \varepsilon_t$ **then**

5 | ⌊ Play arm I_t.

6 | **else**

7 | ⌊ Play a random arm.

Theorem 2.4 (Regret bound for ε_n-greedy [ABF02]) *For all $K > 1$ and for all reward distribution $P_1, \ldots, P_K$ with support in $[0, 1]$, if policy ε_n-greedy is run with input parameter,*

$$0 < d \leq \min_{j:\Delta_j>0} \Delta_j,$$

then the probability that after any number $n \geq cK/d$ of plays ε_n-greedy chooses a suboptimal machine j is at most

$$\frac{c}{d^2n} + 2\left(\frac{c}{d^2}\ln\frac{(n-1)d^2e^{\frac{1}{2}}}{cK}\right)\left(\frac{cK}{(t-1)d^2e^{\frac{1}{2}}}\right)^{c/5d^2} + \frac{4e}{d^2}\left(\frac{cK}{(n-1)d^2e^{\frac{1}{2}}}\right)^{c/2}.$$

$$(2.18)$$

Remark 2.3 To achieve logarithmic regrets in the ε-greedy algorithm, one needs to know the lower bound d on the difference between the reward expectations of the best and the second best machines. One remedy is to replace d^2 with a slowly decreasing term, for instance, $\varepsilon_t \stackrel{def}{=} \min\left\{1, \frac{\ln\ln t}{t}\right\}$. This would result in a regret that grows in the order of $\ln\ln(t) \cdot \ln(t)$, which is (slightly) worse than the bound in (2.18).

Algorithm 2.4: Thompson Sampling for general stochastic bandits

---

1 **Init:** *set* $S_j = 0, F_j = 0$ *for each arm* $j = 1, 2, \ldots, K$.
2 **for** $t = 1, 2, \ldots,$ **do**
3   For each arm $j = 1, \ldots, K$, sample $\theta_j(t)$ from the $Beta(S_j + 1, F_j + 1)$ distribution.
4   Play arm $j(t) := \arg\max_j \theta_j(t)$ and observe reward $\tilde{r}_t$.
5   Perform a Bernoulli trial with success probability $\tilde{r}_t$ and observe output r_t.
6   **if** $r_t = 1$ **then**
7         $\lfloor \ S_{j(t)} = S_{j(t)} + 1$
8   **else**
9         $\lfloor \ F_{j(t)} = F_{j(t)} + 1.$

2.3.3   Thompson Sampling Policy

Thompson sampling (TS) [Tho33] is a natural strategy for MAB problems and has gained a lot of interests recently due to its flexibility to incorporate prior knowledge on the arms. The basic idea is to assume a simple prior distribution on the parameters of the reward distribution of every arm, and at each time step, play an arm according to its posterior probability of being the best arm.

Agrawal and Goyal [AG12] established the first logarithmic regret bound for stochastic MAB using TS. The reward distributions are assumed to have support on [0, 1]. The basic procedure is described in Algorithm 2.4. In Algorithm 2.4, the probability of a successful Bernoulli when arm i is played evolves toward its mean reward. Thus, $\theta_i(t)$ sampled from the Beta distribution would approach the mean reward as arm i is played more times.

The regret bound for K arms using Thompson sampling is given by the following theorem:

Theorem 2.5 (Regret bounds for TS [AG12]) *For the K-armed stochastic bandit problem with support in* [0, 1], *Algorithm 2.4 has an expected regret bound*

$$O\left(\left(\sum_{j:\Delta_j>0} \frac{1}{\Delta_j^2}\right)^2 \ln n\right)$$

in time n. An alternative bound can be obtained as

$$O\left(\frac{\Delta_{max}}{\Delta_{min}^3}\left(\sum_{j:\Delta_j>0} \frac{1}{\Delta_j^2}\right) \ln n\right),$$

where $\Delta_{max} = \max_{j:\Delta_j>0} \Delta_j$ *and* $\Delta_{max} = \min_{j:\Delta_j>0} \Delta_j$.

2.4 Variants of Stochastic Multi-armed Bandit

Many variants of the stochastic MAB have been investigated in literature. In this section, we discuss a few variants that are the most relevant in addressing resource management problems in wireless networks.

2.4.1 Multiplay MAB

In multiplay MAB problems, a player can pull multiple arms in each round. This can be used to model problems, where a decision maker needs to take multiple actions simultaneously. For instance, consider an advertiser who creates an ad campaign to promote her products through a search engine. To participate in search ad auctions, she needs to choose multiple keywords for her campaign. Each keyword can be treated as an arm. As another example, m cooperative users try to access K wireless channels concurrently. The goal is to maximize their sum throughput. In this example, each wireless channel is an arm. If more than two users access the same channel, the reward is zero. Thus, the m channels selected shall be distinct.

One can trivially extend the policies for single-play MAB such as (α, ϕ)-UCB policy or the ε-greedy policy, by viewing each of the $\binom{K}{m}$ possible combinations as an arm. However, doing so ignores the dependence among the rewards from the arms and results in slower learning rate. Recall that the lower bound of the pseudo-regret for Bernoulli bandit in Theorem 2.1 is proportional to $\sum_{j:\Delta_j>0} \frac{\Delta_j}{D(\mu_j\|\mu^*)}$. When the number of arms is large, the lower bound tends to be higher. Though the regret still grows logarithmically, the constant multiplier is bigger.

In [AVW87], Anantharam et. al formulated the multiplay MAB problem, where K arms follow one-parameter reward distributions with the unknown parameter set $(\theta_1, \theta_2, \ldots, \theta_K)$. Successive plays of each arm produce IID rewards. At each stage, one is required to play a fixed number, m, of the arms ($1 \leq m \leq K$). Upon playing m arms, one can observe the reward from each of the arms. An asymptotically efficient allocation rule has been proposed. However, unless the reward distributions are Bernoulli, Poisson, or normal, it is computationally prohibitive to update the statistics required for allocation upon new observation under the proposed allocation rule.

More recently, a Thompson sampling based approach has been proposed for the multiplay MAB problem in [KHN15], where each arm is associated with Bernoulli distribution with unknown means. Similar to the Thompson sampling policy for single-play MAB discussed in Sect. 2.3.3, the success probability θ_j of arm j is assumed to follow a Beta distribution. At time t, having observed $S_j(t)$ successes (reward $= 1$) and $F_j(t)$ failures (reward $= 0$), the posterior distribution of θ_j is updated as $Beta(S_j(t) + 1, F_j(t) + 1)$. The main difference in the multiplay case is to select top m arms ranked by the posterior sample $\theta_j(t)$ from $Beta(S_j(t) + 1, F_j(t) + 1)$.

Komiyama et al. proved that such a simple policy can indeed achieve an any-time logarithmic regret bound.

In [GKJ12], a combinatorial optimization problem with unknown variables and linear rewards has been formulated. Consider a discrete time system with K unknown random processes $X_j(t)$, $1 \leq j \leq K$. $X_j(t)$ is IID with finite support in $[0, 1]$ and mean μ_j. At each decision period t, a K-dimensional action vector $\mathbf{a}(t)$ is selected from a finite set $\mathcal{F}$. Upon selecting an action $\mathbf{a}(t)$, the value of $X_j(t)$ is observed for all j's such that $a_j(t) \neq 0$. The combinatorial optimization problem is thus formulated as

$$\max \quad \sum_t \sum_{j=1}^{K} a_j(t) X_j(t) \tag{2.19}$$
$$s.t. \quad \mathbf{a}(t) \in \mathcal{F}.$$

When the action space $\mathcal{F}$ is restricted to binary vectors in K-dimension with at most m nonzero elements, it is easy to see that the problem is equivalent to the multiplay stochastic MAB problem. Gai et al. proposed a UCB-like policy and proved that it incurs logarithmic regrets over time. At time t, the policy plays an action $\mathbf{a}$ that solves the following maximization problem:

$$\mathbf{a} = \arg\max_{\mathbf{a} \in \mathcal{F}} \sum_{j \in \mathcal{A}_{\mathbf{a}}} a_j \left(\hat{\mu}_j + \sqrt{\frac{(m+1)\ln t}{T_j(t-1)}} \right), \tag{2.20}$$

where $\mathcal{A}_{\mathbf{a}} = \{j \mid a_j \neq 0, j = 1, 2, \ldots, K\}$, $\hat{\mu}_j$ is the sample mean of observed rewards of arm j, and $m = \min\{\max_{\mathbf{a} \in \mathcal{F}} |\mathcal{A}_{\mathbf{a}}|, K\}$ (i.e., the maximum number of arms that can be played in each round). In the special case of the multiplay stochastic MAB, we can see that the above policy plays the top m arms with the highest indices.

2.4.2 MAB with Switching Costs

To this end, our discussion of MAB problems does not consider the costs of switching between arms. In practice, switching costs are important concerns [Jun04]. For example, a worker who switches jobs must pay nonnegligible costs. In job scheduling, switching jobs from one machine to another incurs a variety of setup costs. In wireless networks, changing the operational channels of a radio transmitter incurs nonnegligible delay during which no data can be transmitted. The policies we discuss so far are designed to optimize regrets by playing suboptimal arms less and less often. However, they do not explicitly control the frequency of switching between arms.

Consider K arms whose reward processes are IID with support on $[0, 1]$ and unknown means $\mu_1, \mu_2, \ldots, \mu_K$. Let

$$S_n(j) = \sum_{t=2}^{n} \mathbb{I}_{\{I_{t-1}=j, I_t \neq j\}} \tag{2.21}$$

denote the number of times to switch from arm j to another arm. The switching regret of policy π is then defined as

$$SW_n^\pi = C \sum_{j=1}^{K} \mathbb{E}\left[S_n(j)\right], \qquad (2.22)$$

where $C > 0$ is the fixed switching cost. Therefore, the total regret incurred by a policy π consists two parts, the regret from sampling suboptimal arms, called sampling regret $\overline{R}_n^\pi$ in this context, and the switching regret SW_n^π.

Banks and Sundaram [BS94], and later Sundaram [Sun05] have shown the sub-optimality of index-based policies when switching costs are nonnegligible.[2] In response, researchers have taken three different lines of approaches: characterization of the optimal policy, exact derivation of the optimal policy in restricted settings, and development of order-optimal policies. An excellent survey on this topic can be found in [Jun04]. More recently, Guha and Munagala have investigated the problem, where the switching costs are determined by distances in a metric space [GM09].

Order-optimal policies aim to control the growth of cumulative switching costs such that it is slower than that of the sampling regret from the reward processes. Such policies typically utilize "block" sampling, namely, an action once selected, is in play for a period of time, called an *epoch*. The epoch length is initially short when the uncertainty in the reward processes is high (thus more exploration), and increases as more knowledge is gained so as to amortize the switching costs. Next, we present one such policy based on UCB that was originally proposed by Auer in [ABF02]. Let $\tau(r) = \lceil (1 + \alpha)^r \rceil$, where $\alpha > 0$. Clearly, $\tau(r + 1) - \tau(r) \approx \alpha(1 + \alpha)^r$ grows exponentially with r. We further denote

$$a_{t,r} = \sqrt{\frac{(1 + \alpha) \ln(et/\tau(r))}{2\tau(r)}}. \qquad (2.23)$$

Policy UCB2 in Algorithm 2.5 is an index-based policy and plays arms over epochs that increase exponentially in length over rounds. It has been proven in [ABF02] that policy UCB2 attains logarithmic regrets in time. Furthermore, it is easy to show that when the switching cost is constant between any two arms, the cumulative switching cost incurred is $O(\ln \ln n)$.

2.4.3 Pure Exploration MAB

The last variant of stochastic MAB problems we discuss is the so-called *pure exploration* MAB first studied by Bubeck et al. in [BMS11]. Unlike conventional stochastic MAB problems, where exploitation needs to be performed at the same time as

---

[2]MAB with switching costs can be cast as a restless bandit problem discussed in Chap. 3.

Algorithm 2.5: UCB2 Policy

   Input: $0 < \alpha < 1$.
1 **Init**: *Set $r_j = 0$, for $j = 1, \ldots, K$. Play each machine once.*
2 **for** $t = 1, 2, \ldots,$ **do**
3       $j = \arg\max_{j=1,\ldots,K} \{\hat{\mu}_j + a_{t,r_j}\}$, where $\hat{\mu}_j$ is the average reward obtained from machine j.
4   Play machine j exactly $\tau(r_j + 1) - \tau(r_j)$ times.
5   Set $r_j \leftarrow r_j + 1$

---

exploitation, in pure exploration MAB, the two operate at different stages. A player first selects and observes the rewards of arms for a given number of times n (not necessarily known in advance). She is then asked to provide a recommendation, either deterministic or in the form of a probability distribution over the arms. The player's strategy is evaluated by the difference between the average payoff of the best arm and the average payoff obtained by her recommendation.

Pure exploration MAB is suitable for applications with a preliminary exploration phase in which costs are not measured in terms of rewards but rather in terms of resources consumed. Typically, a limited budget is associated with the resources. For instance, in wireless networks, probing available channels incurs extra delay and energy costs. In formulating channel probing as a pure exploration MAB problem, we aim to design a strategy consisting of a sequence of channels to probe given the delay or energy cost constraints. At the end of the procedure, the "best" channel for data communication is selected and no further probing is needed.

Consider K arms with mean rewards $\mu_1, \mu_2, \ldots, \mu_K$. Denote $P\{1, \ldots, K\}$ the set of all probability distributions over the indexes of the arms. At each round t, a player decides which arm to pull, denoted by I_t, according to a distribution $\varphi_t \in P\{1, \ldots, K\}$ based on past rewards. After pulling the arm, the player gets to see the reward $X_{I_t,T_{I_t}}(t)$, where $T_{I_t}(t)$ is the number of times that arm I_t has been pulled by round t. The primary task is to output at the end of each round t, a recommendation $\psi_t \in P\{1, \ldots, K\}$ to be used to form a randomized play in a one-shot instance if the exploration phase is signaled to be over. The sequence ψ_t is referred to as a recommendation strategy. The *simple regret r_t* of a recommendation $\psi_t = (\psi_{j,t})_{j=1,\ldots,K}$ is defined as the expected regret on a one-shot instance of the game, if a random action is taken according to ψ_t. Formally,

$$r_t = r(\psi_t) = \mu^* - \mu_{\psi_t},$$

where $\mu^* = \max_{j=1,\ldots,K} \mu$ and $\mu_{\psi_t} = \sum_{j=1,\ldots,K} \psi_{j,t}\mu_j$.

From the description of the problem, we see that a player needs to devise two strategies, namely, the allocation strategy φ_t and the recommendation strategy ψ_t. Interestingly, the cumulative expected regret of the allocation strategy is intrinsically related to the simple regret of the recommendation strategy. For Bernoulli reward processes, Bubeck et al. established the following results [BMS11]:

Theorem 2.6 (Simple regret vs. pseudo-regret) *For all allocation strategies (φ_t) and all functions $f : 1, 2, \ldots \to \mathbb{R}$ such that for all (Bernoulli) distributions P_1, $P_2, \ldots, P_K$ on the rewards, there exists a constant $C \geq 0$ with $\mathbb{E}[R_n] \leq Cf(n)$. Then, for all recommendation strategies (ψ_t) based on the allocation strategies (φ_t), there exists a constant $D \geq 0$, such that*

$$\mathbb{E}[r_n] \geq \frac{\min_{j:\Delta_j>0} \Delta_j}{2} e^{-Df(n)}.$$

Theorem 2.6 implies that the smaller the cumulative regret, the larger the simple regret. In [BMS11], several combinations of allocation and recommendation strategies have been analyzed. An immediate consequence of Theorem 2.6 is a lower bound on the simple regret from the fact that the cumulative regrets are always bounded by n:

Corollary 1 *For allocation strategies (φ_t), all recommendation strategies (ψ_t), and all sets of $K \geq 3$ (distinct, Bernoulli) distributions on the rewards, there exist two constants $\beta > 0$ and $\gamma \geq 0$ such that, up to the choice of a good ordering of the considered distributions,*

$$\mathbb{E}[r_n] \geq \beta e^{-\gamma n}. \tag{2.24}$$

The lower bound in (2.24) can be achieved using a trivial uniform allocation policy that samples each arm with equal probability and a recommendation indicating the empirical best arm. However, for moderate values of n, strategies not pulling each arm a linear number of the times in the exploration phase can have interesting simple regrets. One such strategy was discussed in [BMS11], called $UCB(p)$, which is the same as α-UCB policy in Sect. 2.3.1 by setting $\alpha = 2p$. Empirically, it was shown that when the number of arms is large, for moderate n, UCB(p) combined with a recommendation indicating the arm with the highest sample mean or the most played arm incurs smaller simple regrets.

In [AB10], Audibert et al. proposed an elegant algorithm called Successive Reject (SR) for the pure exploration MAB when the number of rounds n of the exploration phase is known. The idea is quite simple. First, the algorithm divides the n rounds into $K - 1$ phases. At the end of each phase, the algorithm dismisses the arm with the lowest sample mean. In next phase, it pulls equally often each arm which has not been dismissed yet. At the end of the n rounds, the last surviving arm is recommended. The length of each phase is chosen carefully such that the simple regret diminishes exponentially with n.

Other variations of the pure exploration problems have been investigated in literature, including the problem of finding the top-m arms [KS10, Kal+12], the multi-bandit problem of finding the best arms simultaneously from several disjoint sets of arms [Gab+11], and the more general combinatorial pure exploration (CPE) problem [Che+14].

2.5 Summary

In this chapter, we have presented the formulation of the stochastic MAB problem and several of its important variants. Representative strategies and their regret bounds have been discussed. Most stochastic MAB problems admit simple index-based policies that can achieve logarithmic regrets in time. It is important to note that the optimality of these index-based policies relies on the assumption of IID reward processes and is expressed in terms of the growth rate in time (as opposed to the number of arms). When the reward process of the arms are dependent, the logarithmic bounds still apply but it is important to exploit the correlation structure among the arms to make learning more efficient.

References

[AB10]   Jean-Yves Audibert and Sébastien Bubeck. "Best arm identification in multi-armed bandits". In: *COLT-23th Conference on Learning Theory*. 2010, 13–p.

[ABF02]   P. Auer, N. C. Bianchi, and P. Fischer. "Finite-time Analysis of the Multiarmed Bandit Problem". In: *Mach. Learn.* 47.2-3 (May 2002), pp. 235–256. ISSN: 0885-6125.

[AG12]   Shipra Agrawal and Navin Goyal. "Analysis of Thompson Sampling for the Multi-armed Bandit Problem." In: 2012.

[AMS09]   Jean-Yves Audibert, Rémi Munos, and Csaba Szepesvári. "Exploration-exploitation Tradeoff Using Variance Estimates in Multi-armed Bandits". In: *Theor. Comput. Sci.* 410.19 (Apr. 2009), pp. 1876–1902. ISSN: 0304-3975.

[AVW87]   V. Anantharam, P. Varaiya, and J. Walrand. "Asymptotically efficient allocation rules for the multiarmed bandit problem with multiple plays-Part I: I.I.D. rewards". In: *Automatic Control, IEEE Transactions on* 32.11 (Nov. 1987), pp. 968–976. ISSN: 0018-9286.

[BC12]   Sébastien Bubeck and Nicol'o Cesa-Bianchi. "Regret Analysis of Stochastic and Nonstochastic Multi-armed Bandit Problems". In: *Foundations and Trends in Machine Learning* 5.1 (2012), pp. 1–122.

[BMS11]   Sébastien Bubeck, Rémi Munos, and Gilles Stoltz. "Pure exploration in finitely-armed and continuous-armed bandits". In: *Theor. Comput. Sci.* 412.19 (2011), pp. 1832–1852.

[BS94]   Jeffrey S Banks and Rangarajan K Sundaram. "Switching costs and the Gittins index". In: *Econometrica* 62.3 (1994), pp. 687–694.

[Che+14] Shouyuan Chen et al. "Combinatorial pure exploration of multi-armed bandits". In: *Advances in Neural Information Processing Systems*. 2014, pp. 379–387.

[Gab+11] Victor Gabillon et al. "Multi-bandit best arm identification". In: *Advances in Neural Information Processing Systems*. 2011, pp. 2222–2230.

[GKJ12]   Y. Gai, B. Krishnamachari, and R. Jain. "Combinatorial Network Optimization with Unknown Variables: Multi-Armed Bandits with Linear Rewards and Individual Observations". In: *IEEE/ACM Transactions on Networking* 20.5 (Oct. 2012), pp. 1466–1478.

[GM09]   Sudipto Guha and Kamesh Munagala. "Multi-armed bandits with metric switching costs". In: *International Colloquium on Automata, Languages, and Programming*. Springer. 2009, pp. 496–507.

[Jun04]   Tackseung Jun. "A survey on the bandit problem with switching costs". In: *De Economist* 152.4 (2004), pp. 513–541.

[Kal+12] Shivaram Kalyanakrishnan et al. "PAC subset selection in stochastic multi-armed bandits". In: *Proceedings of the 29th International Conference on Machine Learning (ICML-12)*. 2012, pp. 655–662.

[KHN15] Junpei Komiyama, Junya Honda, and Hiroshi Nakagawa. "Optimal Regret Analysis of Thompson Sampling in Stochastic Multi-armed Bandit Problem with Multiple Plays". In: *Proceedings of the 32nd International Conference on Machine Learning, ICML 2015, Lille, France, 6-11 July 2015*. 2015, pp. 1152–1161.

[KS10]   Shivaram Kalyanakrishnan and Peter Stone. "Efficient selection of multiple bandit arms: Theory and practice". In: *Proceedings of the 27th International Conference on Machine Learning (ICML-10)*. 2010, pp. 511–518.

[LR85]   T L Lai and H. Robbins. "Asymptotically efficient adaptive allocation rules". In: *Advances in Applied Mathematics* 6.1 (1985), pp. 4–22.

[Sun05]   Rangarajan K Sundaram. "Generalized bandit problems". In: *Social choice and strategic decisions*. Springer, 2005, pp. 131–162.

[Tho33]   William R Thompson. "On the likelihood that one unknown probability exceeds another in view of the evidence of two samples". In: *Biometrika* 25.3/4 (1933), pp. 285–294.

Chapter 3
Markov Multi-armed Bandit

3.1 Problem Formulation

In many application domains, temporal changes in the reward distribution structure are intrinsic characteristics of the problem. For example in wireless networks, changes in channel quality and occupancy are typically modeled as a multistate Markov chain. Markov MAB problems differ from the stochastic MAB problem discussed in Chap. 2 in that the reward process of an arm is not IID, instead, follows a Markov process. Furthermore, the evolution of the Markov processes may depend on the player's decision whether an arm is chosen or not.

Consider K arms indexed by the set $\{1, 2, \ldots, K\}$. The ith arm is modeled as a discrete-time, irreducible, and aperiodic Markov chain with a finite state space S_i. There is a stationary and positive reward associated with each state of each arm. Let r_s^i denote the reward from state s of arm i, $s \in S_i$. Let $P_i = \{p_i(s, s'), s, s' \in S_i\}$ denote the transition probability matrix of the ith arm, and $\pi_i = \{\pi_j(s), s \in S_i\}$ the stationary distribution of P_i. The objective of the Markov MAB problem is to determine the sequence of plays $\{I_1, I_2, \ldots, I_n\}$, where I_t is the arm played at time t such that the cumulative reward is maximized.

In *rested* Markov MAB problems,[1] the state of an arm changes according to P_i if and only if it is played and remains frozen otherwise. Rested Markov MAB can be used to model job scheduling problems in a single server system. The scheduler needs to decide among K jobs, which one to allocate computing resources to at each time interval. The states of the jobs not scheduled remain unchanged. In contrast, in *restless* Markov MAB problems, the state of each arm changes regardless of the user's actions. One application of restless Markov MAB is spectrum sensing and access in cognitive radio networks, which will be discussed in detail in Chap. 5. In spectrum sensing and access, the reward of each arm (channel) is a function of the channel occupancy, which may evolve according to a Markov process independent of the user's action.

---

[1] In some literature, rested Markov MAB is also called sleeping Markov MAB.

© Springer International Publishing AG 2016
R. Zheng and C. Hua, *Sequential Learning and Decision-Making in Wireless Resource Management*, Wireless Networks, DOI 10.1007/978-3-319-50502-2_3

3.1.1   Markov MAB and Markov Decision Process

Markov MAB is intrinsically linked to Markov decision process (MDP) [How60] and reinforcement learning [SB98]. It is important to understand the connections and differences in the formulations.

A MDP is defined by a 5-tuple (S, A, P, R, γ), where S is a finite set of states, A is a finite set of actions, $P_a(s, s') = \mathbb{P}\left(s_{t+1} = s' | s_t = s, a_t = a\right)$ is the probability that action a in state s at time t will lead to state s' at time $t + 1$, $R_a(s, s')$ is the immediate reward (or the expected immediate reward) received after transition to state s' from state s. Let $\gamma \in [0, 1]$ be a discount factor that represents how fast the contribution of the present reward decays in time. The goal of MDP is to find a policy that maximizes the expected discounted rewards over a potentially infinite horizon:

$$\sum_{t=0}^{\infty} \gamma^t R_{a_t}(s_t, s_{t+1}).$$

In MDPs, the transition matrices $P_a(s, s')$ and the rewards $R_a(s, s')$ are known. MDPs can be solved by dynamic programming approaches [Bel57b] such as value iteration [Bel57a] and policy iteration [How60]. The computation complexity of MDPs increases drastically with the sizes of the state space and action space. Important variants of MDP include, partially observable MDP (POMDP), where the system dynamics are determined by an MDP but the underlying states are not directly observable [Son78]; and reinforcement learning, where both the transition probabilities and rewards are not known.

When the transition matrices and the reward distribution are given, a rested Markov MAB can be cast as an MDP. Let $S = \prod_{j=1}^{K} S_j$ and $A = \left\{a \in \{0, 1\}^K | \sum_j^K a_j = 1\right\}$. Denote i_a the index of the nonzero entry of $a \in A$. The transition matrix is given by

$$P_a(s, s') = p_{i_a}(s_{i_a}, s'_{i_a}) \times \prod_{j \neq i_a} \mathbb{I}_{\{s_j = s'_j\}},$$

where $\mathbb{I}_{\{\}}$ is an indicator function. The immediate reward is given by,

$$R_a(s, s') = r_{s_{i_a}}^{i_a}.$$

Rested Markov MAB is fully observable. With the complete information regarding both the (expected) immediate rewards and transition matrices, Gittin first proved the existence of an index-based policy that optimizes discounted cumulative rewards over infinite horizon in [Git79].

In a restless Markov MAB, one has to account for the fact that state transitions occur even for arms that are not selected. In formulating a restless Markov MAB as an MDP, let a state be $(s_j, n_j)_j^K := (s_1, n_1, s_2, n_2, \ldots, s_K, n_K)$. For each arm k, the tuple (s_k, n_k) corresponds to its last observed state and the number of time steps since the last observation of the same arm.

The action space is $\{1, 2, \ldots, K\}$. Let $p_j^l(s, s')$ be the l-step transition probability of the Markov chain underlying arm j. The transition probability from state $(s_j, n_j)_{j=1}^K$ to $(s_j', n_j')_{j=1}^K$ under action j is given by $p_j^{(n_j)}(s_j, s_j')$ iff (i) $n_j' = 1$, (ii) $n_l' = n_l + 1$ and $s_l = s_l'$ for all $l \neq j$. All other transition probabilities are zero. Finally, the mean reward for choosing arm j in state $(s_j, n_j)_{j=1}^K$ is given by $\sum_{s \in S_j} p_j^{(n_j)}(s_j, s) r_s^j$.

Therefore, a restless MAB problem with complete information can in theory be solved using standard techniques for MDP such as value or policy iterations. However, the size of the state space is in fact infinite since $n_j, j = 1, 2, \ldots, K$ can be an arbitrarily large natural number. The decision version of the MDP is known to be among the hardest of problems that can be solved using the amount of memory that is polynomial in the input length. Thus, to have a computational feasible policy, it is important to reduce the size of the state space.

3.1.2 Optimal Policies for Restless Markov MABs with Complete Information

In this section, we present a few examples to gain more insight into the nature of the optimal policies and the difficulties in finding solutions to restless Markov MAB problems. All examples are taken from [Ort+14]. These examples demonstrate that (i) the optimal reward can be (much) bigger than the average reward of the best arm, (ii) the optimal policy does not maximize the immediate reward, and (iii) the optimal policy cannot always be expressed in terms of arm indexes.

Example 1 (Best arm is suboptimal) In this example, the average reward of each of the two arms of a bandit is $\frac{1}{2}$, but the reward of the optimal policy is close to $\frac{3}{4}$. Consider a two-armed bandit. Each arm has two possible states, 0 and 1, which are also the rewards. Underlying each of the two arms is a two-state Markov chain with transmission matrix $\begin{pmatrix} 1 - \varepsilon & \varepsilon \\ \varepsilon & 1 - \varepsilon \end{pmatrix}$, where ε is a small positive quantity. Thus, a typical reward process of each arm goes like

$$00000000001111111111111100000000\ldots.$$

It is easy to see that the optimal policy starts with any arm, and then switches the arm whenever the reward is 0, and otherwise sticks to the same arm. The average reward of the policy is close to $\frac{3}{4}$—much larger than the reward of each arm ($\frac{1}{2}$). Note in this example, making the mean rewards of the arms different by choosing small $\varepsilon_1 \neq \varepsilon_2 > 0$ does not change the nature of the optimal policy.

Example 2 (Another optimal policy) Consider the previous example, but with ε close to 1. A typical reward process of each arm is now

$$0101001011010\ldots$$

Now, the optimal policy switches arms if the reward is 1 and stays otherwise.

Example 1 and 2 also show that the optimal policy is not necessarily a single-action policy. The regret incurred with respect to the optimal single-action policy is often called *weak regret*; on the other hand, *strong regret* is defined as the difference between the rewards attainable by the optimal policy with complete information and the policy under consideration.

Furthermore, we observe from the example, the transition probabilities of the underlying Markov processes (as opposed to mean statistics) play an important role in determining the optimal policy. Though the expected rewards are the identical for both arms in Example 1 and 2, the corresponding policies can be quite different and have different average rewards.

Example 3 (Optimal policy is not myopic) In this example, the optimal policy does not maximize the immediate reward. Again, consider a two-armed bandit. Arm 1 is as in Example 1, and arm 2 provides Bernoulli i.i.d. rewards with probability $\frac{1}{2}$ of getting reward 1. The optimal policy (which knows the distributions) will sample arm 1 until it obtains reward 0, when it switches to arm 2. However, it will sample arm 1 again after some time t (depending on ε), and only switch back to arm 2 when the reward on arm 1 is 0. Note that whatever it is, the expected reward for choosing arm 1 will be strictly smaller than $\frac{1}{2}$, since the last observed reward was 0 and the limiting probability of observing reward 1 is $\frac{1}{2}$. The expected reward of the second arm is always $\frac{1}{2}$. Thus, the optimal policy will sometimes "explore" by pulling the arm with the smaller expected reward.

It was further proven by construction in [Ort+14] that index-based policies are suboptimal.

Theorem 3.1 (Index-based policies are suboptimal) *For each index-based policy π, there is a restless Markov bandit problem in which π behaves suboptimally.*

Despite the negative result, one can seek for computationally efficient algorithms that have sublinear regrets in time when compared to an optimal policy with full knowledge of the transition matrices and immediate rewards.

3.2 Algorithms

In this section, we discuss several algorithms for rested and restless Markov MAB problems.

Algorithm 3.1: Upper confidence bound policy UCB-L

---

1 Init:: $t = 1$ for $t \leq K$ do

2 $\quad \lfloor$ Play arm t; $t = t + 1$

3 while $t > K$ do

4 $\quad \bar{r}_j(T_j(t)) = \frac{r_j(1) + r_j(2) + \dots + r_j(T_j(t))}{T_j(t)}, \forall j$

5 $\quad$ calculate index $g_j = \bar{r}_j(T_j(t)) + \sqrt{\frac{L \ln t}{T_j(t)}}, \forall j$

6 $\quad t = t + 1$

7 $\quad$ Play the arm $I(t) = \arg\max_{j=1,2,\dots,K} g_j$.

8 $\quad \lfloor T_{I(t)}(t+1) = T_{I(t)}(t) + 1$

3.2.1 Rested Markov MAB

In [TL10], Tekin and Liu considered the rested Markov MAB problem with the objective of maximizing the long-term total reward by learning the best arm over time. Here, the optimal policy considered is the best *single-action* policy, which chooses the arm with the highest expected reward. Let $\alpha(t)$ be the arm played by policy α at time t and $s_{\alpha(t)}$ the state of arm $\alpha(t)$ at time t. The (weak) regret of policy α is defined as

$$R^\alpha(n) = n\mu^* - \mathbb{E}\left[\sum_{t=1}^{n} r_{s_{\alpha(t)}}^{\alpha(t)}\right].$$

In the rested MAB problems, since an arm's state does not change when it is not chosen, it suffices to index the reward of an arm by the number of pulls it received so far. In particular, let $r_j(i)$ be the reward of arm j upon the ith pull. It was proven in [TL10] that the α-UCB policy in Sect. 2.3 with parameter $\alpha = 2L$ can in fact achieve logarithmic regrets in time when L satisfies a certain condition. The UCB-L policy is described in Algorithm 3.1. Line 4 updates the sample mean of the respective arm over all possible states. Line 5 computes an index that consists of two parts: sample mean and an estimated upper confidence bound.

Let $\pi_{min} = \min_{s \in S_j, 1 \leq j \leq K} \pi_j(s)$, $r_{max} = \max_{s \in S_j, 1 \leq j \leq K} r_j^s$, $r_{min} = \min_{s \in S_j, 1 \leq j \leq K} r_j^s$, $S_{max} = \max_{1 \leq j \leq K} |S_j|$. Denote ε_j the difference between 1 and the second largest eigenvalue of the transition probability matrix P_j. Let $\varepsilon_{max} = \max_j \varepsilon_j$ and $\varepsilon_{min} = \min_j \varepsilon_j$. Then, the weak regret of the UCB-L policy is stated as follows:

Theorem 3.2 (Regret of UCB-L policy) *Choose* $L \geq 90 S_{max}^2 r_{max}^2 / \varepsilon_{min}$. *The regret of UCB-L policy can be bounded by,*

$$R(n) \leq 4L \sum_{j:\Delta_j > 0} \frac{\ln n}{\Delta_j} + \sum_{j:\Delta_j > 0} \Delta_j \left(1 + (D_j + D_{j*})\beta\right) + \sum_{j=1}^{K} \sum_{s \in S_j} r_j^s \cdot C_{P_j},$$

where C_{P_j} is a constant depending on P_j, $j^* = \arg\max_j \sum_{s \in S_j} \pi_s^j r_s^j$,

$$D_j = \frac{|S_j|}{\pi_{min}} \left(1 + \frac{\varepsilon_{max}\sqrt{L}}{10|S_j|r_{min}} \right),$$

$$\beta = \sum_{t=1}^{\infty} t^{-2} < 2. \tag{3.1}$$

Remark 3.1 The logarithmic regret bound for the UCB-L policy for the rested Markov MAB problem should come as not surprise. The consideration of weak regrets makes the design of an efficient policy essentially an estimation problem. The rested nature ensures that when each time an arm is observed, the state transition is as if the arm is continuously pulled. Therefore, the sample mean converges to the mean reward of an arm. The proof of Theorem 3.2 generally follows that of Theorem 2.2 for the (α, ψ)-UCB policy. The main difference is that one needs to characterize the convergence rate of the sample mean as a function of the number of times an arm is played and each of its states visited.

3.2.2   Restless Markov MAB

3.2.2.1   Single-Action Policies

In restless Markov MAB, since each arm evolves regardless of the player's action, the rewards observed of an arm is a function of the time since the last play of the same arm. The sequence of observations from an arm not played consecutively does not correspond to a discrete-time homogeneous Markov chain. Estimation of the expected rewards from these observations is biased.

To overcome such difficulty, one idea is *to construct a sample path that allows unbiased estimates*. In [TL12], Cekin and Liu proposed an algorithm called *regenerative cycle algorithm* (RCA) and showed that the algorithm incurs logarithmic weak regrets in time. Under RCA, the player maintains a block structure; a block consists of a certain number of slots. Within a block, a player plays the same arm (say, j) continuously till a certain prespecified state (say, γ_j) is observed. Upon this observation, the arm enters a regenerative cycle and the player continues to play the same arm till state γ^j is observed for the second time, which signifies the end of the block. The RCA algorithm singles out for each arm the observations between the first visit and the second visit to state γ_j in each block, virtually assembling them and using them to compute the sample mean and indices. From renewal theory, one can show that the "stitched" process is the same as a sequence of continuous observations from a rested arm statistically.

Denote $SB_1(b)$, $SB_2(b)$, $SB_3(b)$, the three subblocks of block b (Fig. 3.1). The following notations are used in the description of the algorithm.

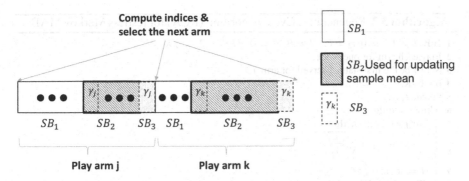

Fig. 3.1 Block structure of RCA. A block is divided into three subblocks SB_1, SB_2, SB_3. The same arm is played in the same block. SB_1 starts from the beginning of the block till an designated state γ is reached. SB_2 spans the slot between the first time and the second time state γ is visited. SB_3 consists of a single slot when γ is visited the second time. At the end of a block, the sample mean and indice of the chosen arm are updated. The arm with the highest index is then selected

- γ_j: the state that determines the regenerative cycles for arm j;
- $b(n)$: the number of completed blocks up to time n;
- $T(n)$: the time at the end of the last completed block;
- $B_j(b)$: the total number of blocks among the first completed b blocks in which arm j is played;
- $X^j(b)$: the vector of observed states from the bth block in which arm is played. Thus, we have $X^j(b) = [X_1^j(b) \ \ X_2^j(b) \ \ \gamma_j]$;
- $t_2(b)$: the number of time slots that lie within an SB_2 till any completed block b;
- $T_2^j(t)$: the number of time slots arm j is played during SB_2's when the number of time steps that lie within an SB_2 is t.

The pseudocode for the RCA algorithm is given in Algorithm 3.2. Comparing Algorithm 3.2 and Algorithm 3.1, we can see the key difference lies in the way the sample mean is updated. In Algorithm 3.1, the sample mean is updated each time an arm is played. In contrast, in Algorithm 3.2, the sample mean of an arm j is updated in the SB_2 subblock of the block that arm j is played. Since the "stitched" process has similar structure as the process produced by a rested arm, a similar regret bound is expected to be held.

The regret bound of Algorithm 3.2 is summarized as follows:

Theorem 3.3 (Regret bound of RCA algorithm [TL12]) *Assume all arms are finite-state, irreducible, aperiodic Markov chains whose transition probability matrices have irreducible multiplicative symmetrizations and the rewards for any arm in any state are positive. Define $\Omega_{max}^j = \max_{s,s' \in S_j} \Omega_{s,s'}^j$, where $\Omega_{s,s'}^j$ is the mean hitting time for arm j to go from state s to state s'. Let $\hat{\pi}_{max} = \max_{s \in S_j, j=1,2,...,K} \{\pi_s^j, 1 - \pi_s^j\}$, $j^* = \arg\max_j \sum_{s \in S_j} \pi_s^j r_s^j$, and $\mu^* = \max_j \sum_{s \in S_j} \pi_s^j r_s^j$.*

With constant $L \geq 112 S_{max}^2 r_{max}^2 \hat{\pi}_{max}^2 / \varepsilon_{min}$, the regret of RCA is upper bounded by,

---

Algorithm 3.2: Regenerative Cycle Algorithm (RCA) for restless Markov MAB

---

1 **Init:** $b = 1, t = 0, t_2 = 0, T_2^j = 0, r^j = 0, \forall j = 1, 2, \ldots, K$
2 Play each arm for one block;
3 Set γ_j be the first state observed for arm j;
4 $b = b+K$;
5 update t, t_2;
6 **while** $t < n$ **do**
7   **for** $j = 1, 2, \ldots, K$ **do**
8          $g^j = \frac{r^j}{T_2^j} + \sqrt{\frac{L \ln t_2}{T_2^j}}$
9      $I = \arg\max_j g^j$
10   Play arm I; Denote observed state as s
11   **while** $s \neq \gamma_I$ **do**
12         $t = t + 1$
13   Play arm I; Denote observed state as s
14     $t = t + 1; t_2 = t_2 + 1; T_2^j = T_2^j + 1; r^I = r^I + r_s(t)$
15   Play arm I; Denote observed state as s
16   **while** $s \neq \gamma_I$ **do**
17         $t = t + 1; t_2 = t_2 + 1; T_2^j = T_2^j + 1; r^I = r^I + r_s(t)$
18   Play arm I; Denote observed state as s
19     $b = b + 1; t = t + 1$

$$R(n) < 4L \ln n \sum_{j:\Delta_j>0} \frac{1}{\Delta_j}(D_j + E_j/\Delta_j) + \sum_{j:\Delta_j>0} C_j(\Delta_j D_j + E_j) + F,$$

where

$$C_j = \left(1 + \frac{(|S_j| + |S_{j*}|)\beta}{\pi_{min}}\right)$$

$$D_j = \left(\frac{1}{\pi_{min}^i} + \Omega_{max}^j + 1\right)$$

$$E_j = \mu_j(1 + \Omega_{max}^j) + \mu^* \Omega_{max}^{j*}$$

$$F = \mu^* \left(\frac{1}{\pi_{min}} + \max_j \Omega_{max}^j + 1\right)$$

$$\beta = \sum_{t=1}^{\infty} t^{-2}. \tag{3.2}$$

Independently, in [LLZ13], Liu et al. proposed a policy based on a deterministic sequencing of exploration and exploitation (DSEE) with an epoch structure. The proposed policy partitions the time horizon into interleaving exploration and exploitation epochs with carefully controlled epoch lengths. An exploration epoch is divided into K contiguous segments, one for playing each of the K arms to learn

---

Algorithm 3.3: Deterministic Sequence Exploration and Exploitation with an Epoch Structure (DSEE)

Input: D
1 **Init**: $n_O = 1; n_I = 0$
2 Play each arm once;
3 $t = K + 1;$
4 **while** $t < n$ **do**
5   **if** $D \log t < (4^{n_O} - 1)/3$ **then**
        /*An exploitation epoch   */
6   Update sample mean $\hat{\mu}_j$ for arm $j = 1, 2, \ldots, K$.
7   Play arm $I = \arg\max_j \hat{\mu}_j$ for 2×4^{n_I} times
8          $n_I = n_I + 1$
9          $t = t + 2 \times 4^{n_I}$
10   **else**
        /*An exploration epoch   */
11   **for** $j = 1, 2, \ldots, K$ **do**
12   Play arm j for 4^{n_O} times
13         $n_O = n_O + 1$
14         $t = t + K \cdot 4^{n_O}$

---

their reward statistics. During an exploitation epoch, the player plays the arm with the largest sample mean (i.e., average reward per play) calculated from the observations obtained so far. The lengths of both the exploration and the exploitation epochs grow geometrically. The pseudocode of the algorithm is given in Algorithm 3.3.

From Algorithm 3.3, in the nth exploration epoch, each arm is played for 4^{n-1} times; whereas in the nth exploitation epoch, the "best" arm with largest sample mean is played $2 \times 4^{n-1}$ times. Furthermore, the frequency of exploration epochs decreases over time (due to the condition in Line 5). In fact, it was proved in [LLZ13] that the number of exploration epochs grows only logarithmically in time. The DSEE algorithm bears some similarity with the ε-greedy algorithm in that in the exploitation phase, only the arm with the highest sample mean (as opposed to an index) is played. However, DSEE utilizes a deterministic epoch structure whereas ε-greedy randomizes the arm to explore. The geometrically growing length of the exploration and exploitation epochs allows unbiased estimation of the sample means.

The weak regret to the DSEE policy is given as follows:

Theorem 3.4 (Regret bound for DSEE [LLZ13]) *Assume all arms are finite-state, irreducible, aperiodic Markov chains whose transition probability matrices have irreducible multiplicative symmetrizations. All rewards are nonnegative. Let $L = \frac{30 r_{max}^2}{(3 - 2\sqrt{2})\varepsilon_{min}}$. Assume the best arm has a distinctive reward mean. The policy parameter D satisfies*

$$D \geq \frac{4L}{(\min_{j:\Delta_j > 0} \Delta_j)^2}.$$

The weak regret of DSEE at time n can be upper bounded by

$$R(n) \leq C_1 \left\lceil \log_4(\frac{3}{2}(n - K) + 1) \right\rceil + C_2[4(3D \ln t + 1) - 1] + K A_{max}(\lfloor \log_4(3D \ln t + 1) \rfloor),$$

where

$$A_{max} = \max_j \left\{ (\min_{s \in S_j} \pi_s^j)^{-1} \sum_{s \in S_j} r_s^j \right\},$$

$$C_1 = \left(3 \sum_{j=1}^{K} K \left(\frac{1}{\ln 2} + \frac{\sqrt{2} \varepsilon_k \sqrt{L}}{10 \sum_{s \in S_j} s} \right) |S_k| \times \sum_{j:\Delta_j>0} \frac{\Delta_j}{\pi_{min}} \right) + A_{max},$$

and

$$C_2 = \frac{1}{3} \left(K \min_{j:\Delta_j>0} \Delta_j - \sum_{j:\Delta_j>0} \Delta_j \right).$$

3.2.2.2 Algorithm to Optimize Strong Regrets

The RCA and DSEE algorithms aim to minimize weak regrets by learning and exploiting the optimal single action. As evident from the examples in Sect. 3.1.2, an optimal policy for a restless Markov MAB problem sometimes needs to switch among the arms. To derive a policy that optimizes strong regrets, one faces challenges in (i) the estimation of the unknown parameters (mean rewards and transition matrices), (ii) the computation complexity to determine the best sequence of actions given the estimated parameters, and (iii) balancing the tradeoff in exploration and exploitation.

In [Ort+14], Ortner et al. introduced the concept of ε-structured MDPs and proposed a general algorithm for learning ε-structured MDPs. It was demonstrated that additional structural information enhances learning. When applied to the restless bandit setting, the algorithm achieves $O(\sqrt{n})$ regret with respect to the best policy that knows the distributions of all arms. The key idea is to aggregate similar states in an MDP so that the complexity of parameter estimation and policy computation can be reduced.

As discussed in Sect. 3.1.1, given the parameters, the restless Markov MAB problem can be cast as an MDP with state space $\{(s_j, n_j)_{j=1}^K\}$. Next, we first introduce the colored UCRL2 algorithm for learning ε-structured MDP and then present the policy for the restless Markov MAB problem. We start with some definitions.

Definition 3.1 (ε-structured MDP) An ε-structured MDP is an MDP with finite state space S, finite action space A, transition probability distribution $p(\cdot|s, a)$, mean reward $r(s, a) \in [0, 1]$, and a coloring function $c : S \times A \to \mathscr{C}$, where $\mathscr{C}$ is the set of colors. Further, for each two pairs $(s, a), (s', a') \in S \times A$ with $c(s, a) = c(s', a')$, there is a bijection function $\phi_{s,a,s',a'} : S \to S$ such that $\sum_{s''} |p(s''|s, a) - p(\phi_{s,a,s',a'}(s'')|s', a')| \leq \varepsilon$ and $|r(s, a) - r(s', a')| < \varepsilon$.

In other words, the state-action pairs in an ε-structured MDP can be grouped into classes, where pairs in the same class have "similar" transition matrices and rewards. The colored UCRL2 algorithm (shown in Algorithm 3.4) was devised for learning in ε-structured MDPs.

Algorithm 3.4: The colored UCLR2 algorithm for learning in ε-structured MDPs

---

Input: Confidence parameter $\delta > 0$, aggregation parameter $\varepsilon > 0$, state space S, action space A, coloring and translation functions, a bound B on the size of the support of transition probability distributions.

1 **Init:** $t = 1$

2 Observe the initial state s_1.

3 **for** *episodes* $k = 1, 2, \ldots$ **do**

4   **Initialize episode k:**

5   Set the start time of episode k, $t_k = t$. Let $N_k(c)$ be the number of times a state-action pair of color c has been visited prior to episode k, and $v_k(c)$ the number of times a state-action pair of color c has been visited in episode k. Compute estimates $\hat{r}_k(s, a)$ and $\hat{p}_k(s'|s, a)$ for rewards and transition probabilities, using all samples from state-action pairs of the same color $c(s, a)$, respectively.

6   **Compute policy** $\tilde{\pi}_k$:

7   Let $\mathcal{M}_k$ be the set of plausible MDPs with reward $\tilde{r}(s, a)$ and transition probabilities $\tilde{p}(\cdot|s, a)$ satisfying

$$|\tilde{r}(s, a) - \hat{r}_k(s, a)| \leq \varepsilon + \sqrt{\frac{7 \log(2Ct_k/\delta)}{2 \max(1, N_k(c(s, a)))}}, \tag{3.3}$$

$$\|\tilde{p}(\cdot|s, a) - \hat{p}_k(s, a)\|_1 \leq \varepsilon + \sqrt{\frac{56B \log(2Ct_k/\delta)}{\max(1, N_k(c(s, a)))}}, \tag{3.4}$$

   where C is the number of distinct colors.

8   Let $\rho(\pi, M)$ be the average reward of a policy $\pi : S \to A$ on an MDP $M \in \mathcal{M}_k$.

9   Choose an optimal policy $\tilde{\pi}_k$ (using value iteration [JOA10]) and an optimistic $\tilde{M}_k \in \mathcal{M}_k$ such that

$$\rho(\tilde{\pi}_k, \tilde{M}_k) = \max\{\rho(\pi, M)|\pi : S \to A, M \in \mathcal{M}_k\}.$$

   Execute policy $\tilde{\pi}_k$:

10   **while** $v_k(c(s_t, \tilde{\pi}_k(s_t))) < \max(1, N_k(c(s_t, \tilde{\pi}_k(s_t))))$ **do**

11   Choose action $a_t = \tilde{\pi}_k$, obtain reward r_t, and observe next state s_{t+1}.

12   Set $t = t + 1$

To convert a restless bandit into an ε-structured MDP, we need to first define the notions of mixing time and diameter. If an arm j is not selected for a large number of time steps, the distribution over states when selecting j will be close to the stationary distribution π_j of the Markov chain underlying arm j. Let π_s^t be the distribution after t steps when starting in state $s \in S_j$. Let

$$d_j(t) = \max_{s \in S_j} ||\pi_s^t - \pi_j||_1 := \max_{s \in S_j} \sum_{s' \in S_j} |\pi_s^t(s') - \pi_j(s')|$$

Definition 3.2 (ε-mixing time) The ε-mixing time of the Markov chain underlying arm j is

$$T_{mix}^j(\varepsilon) := \min\{t \in \mathbb{N} | d_j(t) \leq \varepsilon\}.$$

Definition 3.3 (Diameter of a Markov chain) Let $T_j(s, s')$ be the expected time it takes for the Markov chain underlying arm j to reach state s' from state s, where if $s = s'$, $T_j(s, s') := 1$. The diameter of the Markov chain is defined as $D_j := \max_{s,s' \in S_j} T_j(s, s')$.

To turn any restless bandit into an ε-structured MDP, we assign the same color to two state-action pairs $(s_j, n_j)_{j=1}^K$ and $(s_{j'}, n_{j'})_{j'=1}^K$ iff $j = j'$, $s_j = s_{j'}$ and either $n_j = n_{j'}$ or $n_j = n_j \geq T_{mix}^j(\varepsilon)$. In other words, two states in the MDP are merged if for each arm, they start from the same set of states and either have not been played the same number of times or for sufficiently large number of times (and thus statistically have the same distribution). The respective transition functions are chosen to map states $(s_1, n_1 + 1, \ldots, s_{j-1}, n_{j-1} + 1, s, 1, s_{j+1}, n_{j+1} + 1, \ldots, s_K, n_K + 1)$ to $(s_1', n_1' + 1, \ldots, s_{j-1}', n_{j-1}' + 1, s, 1, s_{j+1}', n_{j+1}' + 1, \ldots, s_K', n_K' + 1)$. This ε-structured MDP can be learned by the colored UCRL2 algorithm introduced earlier with parameter $\varepsilon = 1/\sqrt{T}$. The regret bound is given in Theorem 3.5.

Theorem 3.5 (Regret bound for the colored URCL2 algorithm [Ort+14]) *Consider a restless bandit with K aperiodic arms having state spaces S_j, diameters D_j and mixing time T_{mix}^j, $j = 1, 2, \ldots, K$. Then with probability at least $1 - \delta$ the regret of colored UCRL2 on the transformed ε-structured MDP after $n > 2$ steps is upper bounded by,*

$$90 S \lceil T_{mix} \rceil^{3/2} \prod_{j=1}^K (4D_j) \lceil \max \log_2(4D_i) \rceil \log_2^2 \left(\frac{n}{\delta}\right) \sqrt{n},$$

where $S = \sum_{j=1}^K |S_j|$ is the total number of states and $T_{mix} = \max_{j=1}^K T_{mix}^j$ the maximal mixing time.

Remark 3.2 When comparing the result in Theorem 3.5 with those in Theorems 3.3 and 3.4 for restless MAB, one may be tempted to view it as a weaker result since the regret bound is in square root, rather than logarithmic, in time. However, such a comparison is not fair as Theorem 3.5 characterizes strong regrets. In fact, both Algorithm 3.3 and Algorithm 3.5 have linear strong regrets in time when applies to Example 2 in Sect. 3.1.2. On the other hand, it was proven in [Ort+14] that for any algorithm, there exists a K-armed restless bandit problem with a total number of states $m \times K$, such that the regret after n steps is lower bounded $\Omega(\sqrt{mKn})$. This implies that the colored UCL2 algorithm is order optimal.

Remark 3.3 The colored UCRL2 policy can be applied to rested Markov MAB problems and a similar regret bound applies. Recall that in rested Markov MAB, the states are Cartesian products of the states of the Markov processes underlying individual arms and do not dependent on the time elapse since the last play of the arm. This can be done by applying the following coloring rule: $c((s_j, n_j)_{j=1}^K) = c((s_j', n_j')_{j'=1}^K)$ if $s_j = s_j'$.

3.3 Summary

In this chapter, we discussed the formulations and algorithms for rested and restless MAB problems. Similar to the stochastic MAB problems, variants of the basic formulations have also been considered in literature. Notably, [TL12] examined the optimal single-action policy for multiplay restless and rested MAB problems. Computation aspects of restless MAB problems have been investigated in [WC12].

References

[Bel57a] Richard Bellman. "A Markovian Decision Process". In: *Journal of Mathematics and Mechanics* 6 (1957).

[Bel57b] Richard Bellman. *Dynamic Programming*. 1st ed. Princeton, NJ, USA: Princeton University Press, 1957.

[Git79] John C Gittins. "Bandit processes and dynamic allocation indices". In: *Journal of the Royal Statistical Society. Series B (Methodological)* (1979), pp. 148–177.

[How60] Ronald A. Howard. *Dynamic Programming and Markov Processes*. Technology Press and Wiley, 1960.

[JOA10] Thomas Jaksch, Ronald Ortner, and Peter Auer. "Near-optimal regret bounds for reinforcement learning". In: *Journal of Machine Learning Research* 11.Apr (2010), pp. 1563–1600.

[LLZ13] Haoyang Liu, Keqin Liu, and Qing Zhao. "Learning in a changing world: Restless multiarmed bandit with unknown dynamics". In: *Information Theory, IEEE Transactions on* 59.3 (2013), pp. 1902–1916.

[Ort+14] Ronald Ortner et al. "Regret bounds for restless Markov bandits". In: *Theor. Comput. Sci.* 558 (2014), pp. 62–76.

[SB98] Richard S Sutton and Andrew G Barto. *Reinforcement learning: An introduction*. Vol. 1. 1. MIT press Cambridge, 1998.

[Son78] E.J. Sondik. "The optimal control of partially observable Markov processes over the infinite horizon: discounted cost". In: *Operations Research* 26 (1978), pp. 282–304.

[TL10] Cem Tekin and Mingyan Liu. "Online algorithms for the multi-armed bandit problem with Markovian rewards". In: *Communication, Control, and Computing (Allerton), 2010 48th Annual Allerton Conference on*. IEEE. 2010, pp. 1675–1682.

[TL12] Cem Tekin and Mingyan Liu. "Online learning of rested and restless bandits". In: *IEEE ransactions on Information Theory* 58.8 (2012), pp. 5588–5611.

[WC12] K. Wang and L. Chen. "On Optimality of Myopic Policy for Restless Multi-Armed Bandit Problem: An Axiomatic Approach". In: *IEEE Transactions on Signal Processing* 60.1 (Jan. 2012), pp. 300–309.

Chapter 4
Adversarial Multi-armed Bandit

Abstract In this chapter, we consider the *adversarial MAB* problem, a variant of the MAB problems whereby the stochastic assumption about the processes of rewards is removed. We first introduce the problem and define new notations of regret. We then describe a few well-known algorithms for this problem and provide the asymptotic performance results for these strategies. Finally, we generalize the adversarial MAB problem to the multiplayer case and connect it with the repeated game theory, and show that equilibrium can be achieved with certain strategies.

4.1 Problem Formulation

Adversarial MAB is a class of sequential learning problems whereby a player selects an action I_t among a finite set of K actions at each time step t. Upon making the decision, the player receives a reward $X_{I_t,t} \in (0, 1]$ chosen by an adversary, which depends both on the time instance t and on the chosen action I_t. Unlike classical stochastic MAB problems whereby the rewards are generated randomly and independently following a fixed but unknown distribution, for the adversarial MAB problem, we do not make any statistical assumption about the generation of the reward, which can be generated in either oblivious or nonoblivious opponent models.

Depending on the way the rewards are revealed to the player, this kind of sequential learning problems can be classified into two categories: the *full information game* as shown in Fig. 4.1a whereby the player can get feedback from the adversary about the rewards of all actions, and the *partial information game* as shown in Fig. 4.1b whereby the player only observes the reward of the chosen action, which captures the main feature of the adversarial MAB problem.

In order to characterize the performance of the selected action sequences by this *pure strategy*, the concept of *regret* as defined for the classic stochastic MAB problem in Sect. 2.3 can be adopted to measure the cumulative difference between the reward which could have been achieved if the optimal action is selected, and the reward achieved by playing the actual played actions up to time n. The goal of the player is to minimize its cumulative regret over all possible adversarial assignment of the

© Springer International Publishing AG 2016
R. Zheng and C. Hua, *Sequential Learning and Decision-Making in Wireless Resource Management*, Wireless Networks, DOI 10.1007/978-3-319-50502-2_4

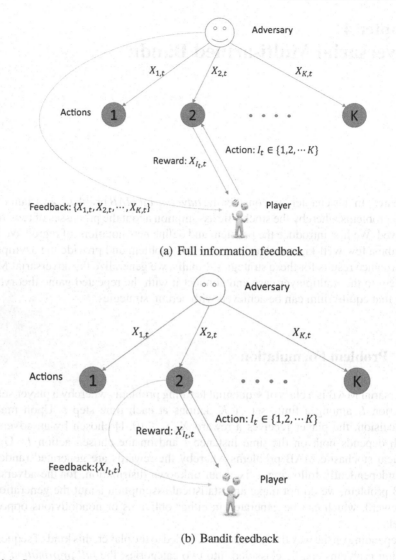

(a) Full information feedback

(b) Bandit feedback

Fig. 4.1 Sequential learning problem with adversary

rewards to all of its actions. However, since the generation process of the rewards is chosen by the adversary, except for some trivial cases, it is impossible to achieve vanishing regret with any deterministic strategy (in particular for the nonoblivious adversary case). For example, consider a toy example with two actions as follows: (i) if $I_t = 1$, then $X_{2,t} = 1$ and $X_{1,t} = 0$; (ii) if $I_t = 2$, then $X_{2,t} = 0$ and $X_{1,t} = 1$. In this case, vanishing regret cannot be achieved since $R_n = n$.

To address this problem, randomization should be introduced into the action selection process, which is known as the *mixed strategy*. That is, instead of selecting a single action at each time instance t, the player chooses a probability distribution $\mathbf{P}_t = (p_{1,t}, \ldots, p_{K,t})$ over the set of K actions, and plays action i with probability $p_{i,t}$. In this case, the notion of *external regret* can be adopted to measure the expected loss if the player chooses the action randomly according to the distribution $\mathbf{P}_t$, defined as follows [CL06]:

Definition 4.1 (*external regret*) For an adversarial MAB problem whereby the player plays with mixed strategies, the cumulative *external regret* is defined as

$$R_n^{ext} = \max_{i=1,\ldots,K} \sum_{t=1}^{n} X_{i,t} - \sum_{t=1}^{n} \bar{X}_{\mathbf{P}_t,t}$$

$$= \max_{i=1,\ldots,K} \sum_{t=1}^{n} \sum_{j=1}^{K} p_{j,t}(X_{i,t} - X_{j,t}), \tag{4.1}$$

where $\bar{X}_{\mathbf{P}_t,t} = \sum_{j=1}^{K} p_{j,t} X_{j,t}$ is the expected reward.

According to the definition, the external regret compares the expected reward of the actual mixed strategy with that of the best action (see Fig. 4.2a). To measure the loss incurred by changing actions in a pairwise manner, another notion of *internal regret* is introduced for the adversarial MAB problem as follows:

Definition 4.2 (*internal regret*) In an adversarial MAB problem whereby the player plays with mixed strategies, the cumulative *internal regret* is defined as

$$R_n^{int} = \max_{i,j=1,\ldots,K} R_{(i,j),n}$$

$$= \max_{i,j=1,\ldots K} \sum_{t=1}^{n} r_{(i,j),t} \tag{4.2}$$

$$= \max_{i,j=1,\ldots K} \sum_{t=1}^{n} p_{i,t}(X_{j,t} - X_{i,t}),$$

where $r_{(i,j),t} = p_{i,t}(X_{j,t} - X_{i,t})$ denotes the expected regret for taking action i instead of action j, which is equivalent to putting the probability mass $p_{i,t}$ on the ith action instead of on the jth one (Fig. 4.2b).

From (4.1) and (4.2), it is easy to show that the external regret is upper bounded by the internal regret as follows:

$$R_n^{ext} = \max_{i=1,\ldots,K} \sum_{j=1,\ldots,K} R_{(i,j),n} \leq \max_{i,j=1,\ldots,K} R_{(i,j),n} = K R_n^{int}, \tag{4.3}$$

which suggests that any algorithm achieving small (vanishing) internal regret also leads to small (vanishing) external regret.

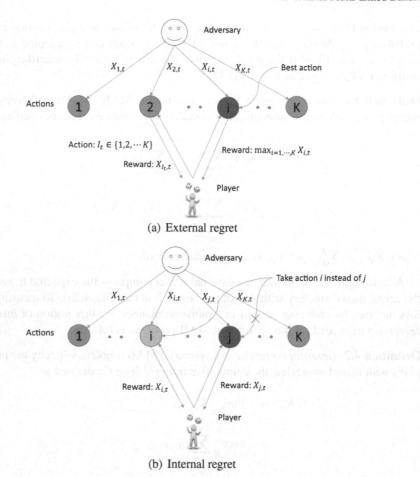

(a) External regret

(b) Internal regret

Fig. 4.2 Examples for regrets: **a** external regret measures the loss of rewards achieved by the best action and the action selected with the mixed strategy; **b** internal regret measures the loss of rewards achieved by choosing action i instead of action j

From the player's perspective, it is desirable to maximize the cumulative expected reward $\sum_{t=1}^{n} X_{I_t, t}$, which can be very large (on the order of n). Instead, we resort to the minimization of regret (external or internal regret). That is, the goal of the player is to select actions so that the per-round regret is vanishing (i.e., negligible as n grows),

$$\lim_{n \to \infty} \frac{1}{n} R_n^{ext} = 0 \quad \text{or} \quad \lim_{n \to \infty} \frac{1}{n} R_n^{int} = 0. \tag{4.4}$$

In other words, $R_n^{ext} = o(n)$ and $R_n^{int} = o(n)$.

4.2   Algorithms

In this section, we discuss some representative mixed strategies that can achieve vanishing regrets. These strategies are based on the randomized prediction scheme, which differ from each other in the way that the selection probabilities of actions are updated in each time instance.

4.2.1   Weighted Average Prediction Algorithm

The basic idea of the *weighted average prediction* (WAP) strategy is to compute the probability for choosing an action i as follows:

$$p_{i,t} = \frac{w_{i,t-1}}{\sum_{j=1}^{K} w_{j,t-1}}, \tag{4.5}$$

where $w_{1,t-1}, \dots, w_{K,t-1} \geq 0$ are the weights assigned to the actions at time t. In order to minimize that regret, the weights are chosen according to the regret up to time $t - 1$, that is, a large $w_{i,t-1}$ is assigned to action i if its cumulative regret $\sum_{s=1}^{t-1} X_{i,s} - \sum_{s=1}^{t-1} \bar{X}_{\mathbf{P}_s,s}$ is large. Therefore, this strategy tends to weight more those actions i whose cumulative reward $\sum_{s=1}^{t-1} X_{i,s}$ is large.

Hedge algorithm [Aue+95] is one of the WAP algorithms for the full information game as shown in Fig. 4.1a, i.e., the rewards of all actions can be observed after each decision, which is a variant of the *Weighted Majority* algorithm [LW89]. As shown in Algorithm 4.1, **Hedge** algorithm adopts the exponentially weighted average (EWA) strategy, which selects the ith action at time instance t with probability proportional to $exp(\eta \sum_{s=1}^{t-1} X_{i,s})$, specifically,

$$p_{i,t} = \frac{\exp(\eta \sum_{s=1}^{t-1} X_{i,s})}{\sum_{j=1}^{K} \exp(\eta \sum_{s=1}^{t-1} X_{j,s})}, \tag{4.6}$$

Algorithm 4.1: Hedge algorithm

---
1 **Init:** $\eta > 0$;
2 **for** $t = 1, 2, \dots$ **do**
3   **for** $i = 1, \dots, K$ **do**
4   Set probability $p_{i,t} = \frac{\exp(\eta \sum_{s=1}^{t-1} X_{i,s})}{\sum_{j=1}^{K} \exp(\eta \sum_{s=1}^{t-1} X_{j,s})}$;
5   Select action I_t according to the distribution $P_t = (p_{1,t}, \dots, p_{K,t})$;
6   Play action I_t and receive reward $X_{i,t}$ for all $i = 1, \dots, K$;

The following theorem provides the upper bound for the external regret of the **Hedge** algorithm:

Theorem 4.1 (External regret of Hedge) *For any n and $\eta > 0$, the external regret of the **Hedge** algorithm is upper bounded by*

$$R_n^{ext} \leq \frac{\ln K}{\eta} + \frac{n\eta}{8}. \tag{4.7}$$

Proof Let $w_{i,t} = \exp(\eta \sum_{s=1}^{t-1} X_{i,s})$, $W_t = \sum_{i=1}^{N} w_{i,t}$, then $p_{i,t} = \frac{w_{i,t}}{W_t}$.
   We have

$$\ln \frac{W_{n+1}}{W_1} = \ln \left(\frac{1}{K} \sum_{i=1}^{K} w_{i,n+1} \right)$$

$$\geq \ln \left(\frac{1}{K} \max_i w_{i,n+1} \right) \tag{4.8}$$

$$\geq \eta \max_i \sum_{t=1}^{n} X_{i,t} - \ln K.$$

On the other hand,

$$\ln \frac{W_{n+1}}{W_1} = \sum_{t=1}^{n} \ln \frac{W_{t+1}}{W_t}$$

$$= \sum_{t=1}^{n} \ln \left(\sum_{i=1}^{K} \frac{w_{i,t}}{W_t} \exp(\eta X_{i,t}) \right)$$

$$= \sum_{t=1}^{n} \ln(\mathbb{E} \exp(\eta X_{I_t,t})) \tag{4.9}$$

$$\leq \sum_{t=1}^{n} \left(\eta \mathbb{E} X_{I_t,t} + \frac{\eta^2}{8} \right),$$

where the last inequality is obtained by the classic inequality [Hoe63]

$$\ln \mathbb{E} \exp(sX) \leq s\mathbb{E}X + \frac{s^2}{8}$$

for a random variable $X \in [0, 1]$ and $s \in \mathbb{R}$.
   By combining (4.8) and (4.9), we obtain the inequality in Theorem 4.1:

$$\max_{i=1,\dots,K} \sum_{s=1}^{n} X_{i,s} - \sum_{s=1}^{n} \mathbb{E} X_{I_s,s} \leq \frac{n\eta}{8} + \frac{\ln K}{\eta}.$$

If the time horizon n is known, with the choice $\eta = \sqrt{8 \ln K / n}$, the external regret is upper bounded by $R_n^{ext} \leq \sqrt{n \ln K / 2}$. Otherwise, if the horizon is unknown, with a time-varying parameter $\eta_t = \sqrt{8 \ln K / t}$, the upper bound of the external regret becomes $R_n^{ext} \leq 2\sqrt{n \ln K / 2} + \sqrt{\ln K / 8}$.

Algorithm 4.2: EXP3 algorithm

Input: $\gamma \in (0, 1], w_{i,1} = 1, i = 1, \ldots, K$

1 **for** $t = 1, 2, \ldots$ **do**

2   **for** $i = 1, \ldots, K$ **do**

3   Set $p_{i,t} = (1 - \gamma)\dfrac{w_{i,t}}{\sum_{j=1}^{K} w_{j,t}} + \dfrac{\gamma}{K}$;

4   Draw action I_t randomly according to the probabilities $p_{1,t}, \ldots, p_{K,t}$;

5   Play action I_t and receive reward $X_{I_t,t}$;

6   **for** $j = 1, \ldots, K$ **do**

7   Set

$$\hat{X}_{j,t} = \begin{cases} \dfrac{X_{j,t}}{P_{j,t}} & \text{if } j = I_t \\ 0, & \text{otherwise,} \end{cases} \qquad (4.10)$$

      and

$$w_{j,t+1} = w_{j,t} \exp\left(\dfrac{\gamma \hat{X}_{j,t}}{K}\right)$$

EXP3 algorithm is an extension of the **Hedge** algorithm for the adversarial MAB problem with partial information as shown in Fig. 4.1b [Aue+02], that is, only the reward of the selected action can be observed. As shown in Algorithm 4.2, **EXP3** selects an action I_t according to the distribution $\mathbf{P}_t$ in each time step t, which is a mixture of the probability $\mathbf{P}_t$ defined in the Hedge algorithm and the uniform distribution to ensure that all actions are tried and the reward for each action is estimated.

Upon receiving the reward for the chosen action I_t, **EXP3** sets the estimated reward $\hat{X}_{I_t,t} = X_{I_t,t}/p_{I_t,t}$, which is an unbiased estimation of $X_{I_t,t}$ at time t for all $j = 1, \ldots, K$ since $\mathbb{E}[\hat{X}_{j,t}|I_1, \ldots, I_t] = X_{j,t}$.

The following theorem provides the bound for the *external regret* of **EXP3**:

Theorem 4.2 (External regret of EXP3, [Aue+02]) *For any $n > 0$, assume that* $g \geq \max_j \sum_{t=1}^{n} X_{j,t}$, *with the parameter of*

$$\gamma = \min\left\{1, \sqrt{\frac{K \ln K}{(e - 1)g}}\right\}, \qquad (4.11)$$

then the expected external regret of **EXP3** *is given by*

$$R^{ext} \leq 2\sqrt{e - 1}\sqrt{gK \ln K} \qquad (4.12)$$

This upper bound is obtained under the condition that g is known in advance, and thus the parameter γ can be set according to (4.11). In particular, if the time horizon n is known, then g can be set to n since the reward for each trial is at most 1, so the expected external regret is upper bounded by $2\sqrt{e - 1}\sqrt{nK \ln K}$.

Theorem 4.2 provides the *expected* regret of EXP3, but the variance of the regret achieved by this algorithm can be very large due to the large variance of the estimate $\hat{X}_{j,t}$ for the reward $X_{j,t}$. In order to control the variance, a new strategy can be adopted to estimate the reward as follows:

$$X'_{j,t} = \hat{X}_{j,t} + \frac{\beta}{p_{j,t}} = \begin{cases} \dfrac{X_{j,t} + \beta}{p_{j,t}}, & \text{if } j = I_t \\ \dfrac{\beta}{p_{j,t}}, & \text{otherwise,} \end{cases} \tag{4.13}$$

where β is a positive parameter.

Note that $X'_{j,t}$ is no longer an *unbiased estimation* of the reward, but a little larger than $\hat{X}_{j,t}$. This guarantees that the estimated reward is not much smaller than the actual (unknown) cumulative rewards with high probability, so it can be interpreted as an upper confidence bound on the reward.

The regret of the modified EXP3 algorithm (by changing (4.10) in Algorithm 4.2 with (4.13)) is given by the following theorem:

Theorem 4.3 (High probability bound for the external regret of the modified EXP3, [CL06]) *For any $\delta \in (0, 1)$ and $n \geq 8K \ln(K/\delta)$, with the setting of*

$$\beta = \sqrt{\frac{1}{nK} \ln \frac{K}{\delta}}, \gamma = \frac{4K\beta}{3 + \beta}, \text{and } \eta = \frac{\gamma}{2K}$$

then the external regret is upper bounded by

$$R_n^{ext} \leq \frac{11}{2}\sqrt{nK \ln(K/\delta)} + \frac{\ln K}{2} \tag{4.14}$$

with probability at least $1 - \delta$.

4.2.2   Following-the-Perturbed-Leader (FPL) Algorithm

FPL algorithm [KE07] is an online prediction scheme similar to the weighted average strategy, which is an extension of the basic *following-the-leader* strategy [Han57] for the full information game. In each time step, the player selects the action with the maximum accumulative reward in the past. To achieve vanishing regrets, the FPL algorithm adds a random *perturbation* to the accumulated reward of each action, such that the action that achieves the maximum perturbed reward is selected at each time instance.

Specifically, let $Z_{i,t}(i = 1, \ldots, K)$ denote the random perturbation, then the FPL algorithm selects the action with the highest perturbed rewards at each time instance t, i.e.,

$$I_t = \arg \max_{i=1,\ldots,K} \left(\sum_{s=1}^{t-1} X_{i,s} + Z_{i,t} \right). \tag{4.15}$$

Depending on the distribution of the perturbations, different upper bounds can be achieved for the external regret of the FPL algorithm.

Theorem 4.4 (External regret of FPL with uniformly distributed perturbations, [CL06]) *If $Z_{i,t}$s are independent with uniform distribution on $[0, \Delta]$, then by the choice of $\Delta = \sqrt{nK}$, the external regret of the FPL algorithm with an oblivious opponent satisfies*

$$R_n^{ext} \leq 2\sqrt{nK}. \tag{4.16}$$

Moreover, with any (nonoblivious) opponent, with probability at least $1 - \delta$, the external regret satisfies

$$R_n^{ext} \leq 2\sqrt{nK} + \sqrt{\frac{n}{2} \ln \frac{1}{\delta}}. \tag{4.17}$$

From Theorem 4.4, it can be seen that the upper bound of the regret is on the order of $O(\sqrt{nK})$, which is much worse than that achieved by the Hedge algorithm. To this end, the distribution of $Z_{i,t}$ should be chosen carefully. For example, $Z_{i,t}$ can be generated according to the *two-side exponential distribution* of parameter $\eta > 0$ so that the joint density of Z_t is $f(Z) = (\eta/2)^K e^{-\eta \|Z\|_1}$, where $\|Z\|_1 = \sum_{i=1}^{K} Z_{i,t}$. Then we have the following result regarding the external regret of the FPL algorithm:

Theorem 4.5 (External regret of FPL with two-side exponential distributed perturbations[CL06], Remark 4.2) *If $Z_{i,t}$ are independent with two-side exponential distribution of parameter $\eta > 0$, then with the choice of $\eta = \min\{1, \sqrt{2(1 + \ln K)/((e - 1)n)}\}$, the external regret under any nonoblivious opponent is upper bounded with probability at least $1 - \delta$ as follows:*

$$R_n^{ext} \leq 2\sqrt{2(e-1)n(1 + \ln K)} + 2(e+1)(1 + \ln K) + \sqrt{\frac{n}{2} \ln \frac{K}{\delta}}, \tag{4.18}$$

This upper bound is comparable to that achieved by the weighted average prediction strategies such as Hedge.

An adaptive bandit FPL (AB-FPL) algorithm is proposed for the adversarial MAB problem in [KE07]. As shown in Algorithm 4.3, the algorithm uses $\hat{X}'_{j,t} = \hat{X}_{j,t} - \lambda \sigma_{j,t} \sqrt{\ln(t + 1)}$ as an estimate of the reward, where λ is a small constant, and $\sigma_{j,t}$ is the upper bound of the conditional standard deviation of $\hat{X}_{j,t}$. Then in each time step, the algorithm selects an action between exploration and exploitation, namely, with

---

Algorithm 4.3: AB-FPL algorithm

---

1 Init: ϵ_t : *the width of the perturbation vector* μ_t *at the exploitation stage. For known horizon*
  n, *set* $\epsilon_t = \frac{1}{3}\sqrt{\frac{\ln n}{nK}}$.
2 $\gamma_t = \min(1, K\epsilon_t)$: *the probability of selecting an action uniformly at random;*
3 for $t = 1, 2, \cdots$ **do**
4   Compute the unbiased estimation of the reward

$$\hat{X}_{j,t} = \begin{cases} \dfrac{X_{j,t}}{p_{j,t}}, & \text{if j is selected at time } t; \\ 0, & \text{otherwise.} \end{cases} \tag{4.19}$$

Compute the upper bound for the conditional variance of $\hat{X}_{j,t}$ as follows:

$$\sigma_{t,j}^2 = \sum_{s=1}^{t} \frac{1}{p_{j,s}} \geq \sum_{s=1}^{t} \frac{X_{j,t}^2}{p_{j,s}} - X_{j,s}^2 = \sum_{s=1}^{t} Var(\hat{X}_{j,s}|s-1,\cdots,1). \tag{4.20}$$

Define $\hat{X}'_{j,t} = \hat{X}_{j,t} - \lambda\sigma_{j,t}\sqrt{\ln(t+1)}$, where $\lambda = \sqrt{1+\sqrt{2/k}}$;
5 (**Exploration stage**) : Select an action uniformly from K actions with probability γ_t;
6 (**Exploitation stage**): Otherwise, select the action I_t with the highest perturbed reward

$$I_t = \arg\max_{i=1,\cdots,K}(\sum_{s=1}^{t-1}\hat{X}'_{i,t} + \mu_{i,t}), \tag{4.21}$$

where $\mu_{i,t} \propto \exp(-\epsilon_t\|x\|_1)$ is a random perturbation drawn from the two-sided
exponential distribution.

---

probability γ_t, it selects an action uniformly from all actions to explore the reward, and
with probability $1 - \gamma_t$, it selects the action which achieves the maximum perturbed
reward.

Theorem 4.6 (External regret of AB-FPL) *The external regret of the AB-FPL algorithm satisfies*

$$R_n^{ext} \leq 5\sqrt{nK\ln n}. \tag{4.22}$$

Note that the computation of the estimated reward in (4.19) relies on the probability
$p_{j,t}$, which is not explicitly defined in the AB-FPL algorithm. However, according to
the action selection policy, the value of $p_{j,t}$ only depends on the joint distribution of
the random variables $Z_{j,t}$ but not their random values. Since $Z_{j,t}$ is i.i.d. following
two-sided exponential distribution with the expected value $\sum_{s=1}^{t-1}\hat{X}'(j,s)$, $p_{j,t}$ is the
probability that action j has the highest value. That is, $p_{j,t} = \int_{-\infty}^{\infty} p_1(x)p_2(x)dx$,
where $p_1(x)$ is the density of the value for action j, $p_2(x)$ is the probability that
the values of other actions are lower than that of action j, which is a product of the
cumulative distributions for the exponential distribution of all other $K-1$ actions.

4.2.3 Implicitly Normalized Forecaster (INF) Algorithm

INF algorithm [AB09, AB10] is a new class of randomized policies, which is based on a potential function $\psi : \mathbb{R}_*^- \to \mathbb{R}_+$, which is increasing, convex, twice continuously differentiable, and ψ'/ψ is nondecreasing, and

$$\lim_{x \to -\infty} \psi(x) < 1/K, \text{ and } \lim_{x \to 0} \psi(x) \geq 1.$$

For example, if $\psi(x) = \exp(\eta x) + \frac{\gamma}{K}$, INF corresponds exactly to the EXP3 strategy. If $\psi(x) = (\frac{-\eta}{x})^q + \frac{\gamma}{K}$ with $q > 1$ and $\eta > 0$, it is called PolyINF strategy.

Algorithm 4.4: INF algorithm

---

1 **Init:** *Set p_1 to be the uniform distribution over $\{1, \cdots, K\}$;*
2 **for** $t = 1, 2, \cdots$ **do**
3   Select an action I_t according to probability distribution p_t;
4   Play action I_t and receive reward $X_{I_t,t}$;
5   **for** $i = 1, \cdots, K$ **do**
6   Compute the estimated reward

$$\hat{X}_{i,t} = \begin{cases} \dfrac{X_{i,t}}{p_{i,t}}, \text{ if } i = I_t, \\ 0, \text{ otherwise.} \end{cases} \qquad (4.23)$$

7   **for** $i = 1, \cdots, K$ **do**
8   Compute the new probability distribution

$$p_{i,t} = \psi\left(\sum_{s=1}^{t-1} \hat{X}_{i,t} - C_t\right), \qquad (4.24)$$

        where C_t is the unique real number such that $\sum_{i=1}^{K} p_{i,t} = 1$.

As shown in Algorithm 4.4, at each time step t, INF selects an action according to a probability distribution defined as $p_{i,t} = \psi\left(\sum_{s=1}^{t-1} \hat{X}_{i,t} - C_t\right)$, where C_t is the unique real number such that $\sum_{i=1}^{K} p_{i,t} = 1$. Therefore, the probability is proportional to the function of the difference between the estimated cumulative reward of each action and the cumulative reward of the policy, which is similar to traditional weighted average prediction strategies such as EXP3.

We have the following results about the external regret for the INF algorithm.

Theorem 4.7 (External regret of INF, [AB10]) *For the INF strategy with* $\psi(x) = \exp(\eta x) + \frac{\gamma}{K}$, *where* $0 < \frac{4\eta K}{5} \leq \gamma < 1$, *the external regret satisfies*

$$R_n^{ext} \leq \sqrt{\frac{16}{5} n K \ln K}.$$

Theorem 4.8 (External regret of PolyINF, [AB10]) *For the PolyINF strategy with* $\psi(x) = (\frac{-\eta}{x})^q + \frac{\gamma}{K}$, *where* $q = 2, \eta = \sqrt{5t}$, *and* $\gamma = \min(1/2, \sqrt{3K/n})$, *the external regret satisfies*

$$R_n^{ext} \leq 8\sqrt{2nK},$$

From Theorems 4.7 and 4.8, it can be seen that the PolyINF strategy achieves better regret bound since it removes the extraneous $\ln K$ factor.

4.2.4 Internal-Regret Minimizing Algorithm

According to (4.3), any algorithm achieving small internal regret also has a small external regret, but the converse is not necessarily true. Indeed, as we have shown in previous subsections, the randomized prediction algorithms such as EXP3, FPL, and INF achieve vanishing external regret (i.e., sublinear in n), but these algorithms may have a linearly growing internal regret.

On the other hand, achieving vanishing internal regret is important due to its tight connection to the correlated equilibrium (to be discussed in next section). To address this issue, a simple method is proposed in [SL05] to convert an external-regret minimizing algorithm into an internal-regret minimizing one. The basic procedure of the method is shown in Algorithm 4.5. At each time step t, the player selects an action according to a probability distribution $\mathbf{P}_t$, which may be started with uniform distribution, i.e., $P_1 = (1/K, \ldots, 1/K)$. Based on P_t, we can define $K(K-1)$ virtual actions indexed by a pair of integers $i \neq j$. Each pair (i, j) of action is associated with a probability distribution $\mathbf{P}_t^{i \to j}$, which is obtained from $\mathbf{P}_{t-1}$ using the $i \to j$ *modified strategy*, that is, the probability mass is transferred from i to j, so the ith component of $\mathbf{P}_t^{i \to j}$ is zero, and the jth component equals to $p_{i,t-1} + p_{j,t-1}$, while the remaining components are the same as those of $\mathbf{P}_t$, so $\mathbf{P}_t^{i \to j} = (p_{1,t}, \ldots, 0, \ldots, p_{i,t-1} + p_{j,t-1}, \ldots, p_{K,t-1})$. In addition, the expected reward corresponding to the modified strategy is given by $\bar{X}_{\mathbf{P}_t^{i \to j}, t} = \sum_{i=1}^{K} P_{i,t}^{i \to j} X_{i,t}$.

Suppose that an external-regret minimizing algorithm is adopted for these $K(K-1)$ virtual actions, and let $\Delta_{i \to j, t}$ denote the obtained probability distribution for the pair (i, j) of action, then $\mathbf{P}_t$ can be obtained by solving the following fixed point equation:

$$\mathbf{P}_t = \sum_{(i,j):i \neq j} \mathbf{P}_t^{i \to j} \Delta_{i \to j, t}. \tag{4.25}$$

It is proven in [FV99, SL05] that such $\mathbf{P}_t$ exists and can be computed using the Gaussian elimination method.

Algorithm 4.5: Internal-regret Minimizing Algorithm

---

1 **Init**: $P_1 = (\frac{1}{K}, \cdots, \frac{1}{K})$;
2 Select an action using P_1;
3 Play the action and observe the reward;
4 **for** $t = 2, \cdots,$ **do**
5   **for** $i, j = 1, \cdots, K, i \neq j$ **do**
6   Construct $P_t^{i \to j}$ using the $i \to j$ *modified strategy*, that is,

$$\mathbf{P}_t^{i \to j} = (p_{1,t-1}, \cdots, 0, \cdots, p_{i,t-1} + p_{j,t-1}, \cdots, p_{K,t-1})$$

7   Compute the expected reward for the modified strategy:

$$\bar{X}_{\mathbf{P}_t^{i \to j},t} = \sum_{i=1}^{K} P_{i,t}^{i \to j} X_{i,t}$$

8   Update $\Delta_{(i,j),t}$ using any external-regret minimizing algorithm;
9   Find P_t by solving the following fixed point equation:

$$\mathbf{P}_t = \sum_{(i,j):i \neq j} \mathbf{P}_t^{i \to j} \Delta_{i \to j,t}.$$

   Select an action using $\mathbf{P}_t$ and observe the reward.

This procedure can be applied to convert any external-regret minimizing algorithm into internal-regret minimizing one. In particular, we have the following property about the internal regret of this procedure based on the exponentially weighted average (EWA) strategy:

Theorem 4.9 (Internal regret based on EWA strategy, [SL05]) *If Δ_t is obtained using the exponentially weighted average strategy, i.e., Δ_t is given by*

$$\Delta_{(i,j),t} = \frac{\exp(\eta \sum_{s=1}^{t-1} \bar{X}_{\mathbf{P}_s^{i \to j},s})}{\sum_{(k,l):k \neq l} \exp(\eta \sum_{s=1}^{t-1} \bar{X}_{\mathbf{P}_s^{k \to l},s})}, \tag{4.26}$$

then with the choice of $\eta = 4\sqrt{\ln K / n}$ for known time horizon n, the internal regret satisfies

$$R_n^{int} = \max_{i \neq j} r_{(i,j),n} \leq \sqrt{n \ln K}. \tag{4.27}$$

4.3   Game Theoretical Results for Multiplayer Adversarial Multi-armed Bandit

The adversarial MAB framework can be generalized to the multiplayer case (Fig. 4.3). Specifically, at each time instance t, each player $j \in \{1, \ldots, N\}$ selects an action $I_t^{(j)}$ from its set of actions $\{1, \ldots, K_j\}$ according to a mixed strategy given by the probability distribution $\mathbf{p}_t^{(j)} = (p_{1,t}^{(j)}, \ldots, p_{K_j,t}^{(j)})$. After playing the action, each player only observes its own reward of the selected action, which is not revealed to other players. The reward of each player depends not only on its selection, but also on the decisions of other players. More formally, let $\mathbf{I}_t = (I_t^{(1)}, \ldots I_t^{(N)})$ denote the joint action profile selected by all players at time t, where $I_t^{(j)} \in \{1, \ldots, K_j\}$ is the action selected by player j, then the reward received by player j is $X_t^{(j)}(\mathbf{I})$: $\bigotimes_{j=1}^{N}\{1, \ldots, K_j\} \rightarrow [0, 1]$.

From the game theoretical perspective, this multiplayer adversarial MAB problem can be modeled as an N-player repeated game, whereby a player j is competing with other $N-1$ opponents at each time instance by repeating the aforementioned mixed strategy. In particular, we assume that the playing of all players is "uncoupled", that is, each player adopts a *regret-based* procedure to update its mixed strategy, which depends only on its past rewards. In this case, it is interesting to study whether this repeated game can converge to some kinds of equilibrium. Indeed, for the special case of two-person zero-sum games, the convergence to Nash equilibrium can be achieved if both players adopt an external-regret minimizing strategy (e.g., EXP3, AB-FPL, etc.). However, this property does not hold for general N-player game

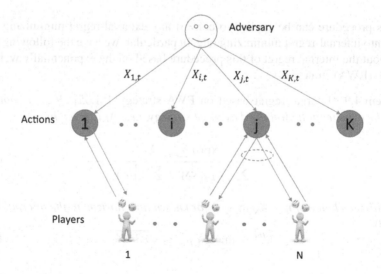

Fig. 4.3 Multiplayer adversarial MAB problem

with $N \geq 3$, that is, if players adopt external-regret minimizing strategy, it is not guaranteed to approach the Nash equilibrium or the correlated equilibrium.

On the other hand, if each player plays according to an internal-regret minimizing strategy (which may be converted from any external-regret minimizing strategy using the method introduced in Sect. 4.2.4), the joint empirical frequencies of plays converge to a *correlated equilibrium* as shown in the following theorem.

Theorem 4.10 (Convergence of the internal-regret minimizing strategy to the correlated equilibrium, [CL06]) *Consider a repeated game with N players, and its set of correlated equilibria is denoted by C. If each player plays according to an internal-regret minimizing strategy, then the distance $\inf_{p \in C} \sum_i \| P(i) - \hat{P}_t(i) \|$ between the empirical distribution of plays and the set of correlated equilibria converge to 0 almost surely.*

That is, if an internal-regret minimizing strategy (i.e., it yields vanishing internal regret) is adopted by all players, then the convergence of the game to the correlated equilibrium is guaranteed almost surely. This result can be extended to the *unknown games*, which consist of two agents from the view point of a player k: the first agent is player k itself, and the second agent is the set of all other $K - 1$ players whose actions affect the reward achieved by player k. In this case, each player does not know the total number of players as well as the reward function of any player (including itself), and after taking an action at each time instance t, each player only sees its own reward, but the choices of other players and their rewards are not revealed to the player. In this case, each player can play according to some internal-regret minimizing strategies, which guarantees that the empirical frequencies of play will converge to the set of correlated equilibria of the game.

For this unknown game setup, it is also possible to achieve an approximate Nash equilibrium using the experimental regret testing (ERT) algorithm [CL06]. As shown in Algorithm 4.6, the time is divided into periods of length $T > 0$. At the end of each period, the player calculates its experimental regret, if it is larger than a prescribed threshold, a new mixed strategy is selected randomly and played in the next period. Otherwise, it continues using current mixed strategy in the next period.

The key of the ERT algorithm is the calculation of the experimental regret. To this end, let us consider a period between $(m - 1)T + 1$ and mT and define a random variable $U_{k,s} \in \{0, 1, \ldots, K_j\}$ for $s \in [(m - 1)T + 1, mT]$. For this period, player j draw n_j samples for $i_j = 1, \ldots, K_j$, then there are n_j values of s such that $U_{j,s} = i_j$, and for the remaining s, $U_{j,s} = 0$. At time instance s, the action $I_s^{(j)}$ is selected as follows:

$$I_s^{(j)} : \begin{cases} \text{is distributed as } P_{m-1}^{(j)} & \text{if } U_{j,s} = 0 \\ \text{equals } i_j & \text{if } U_{j,s} = i_j. \end{cases}$$

Algorithm 4.6: ERT Algorithm for player j

---

1 **Init:** *Period length T, confidence parameter $\rho > 0$, exploration parameter $\lambda > 0$;*
2 Choose a mixed strategy $P_0^{(j)}$ randomly according to the uniform distribution over $\{1, \cdots, K_j\}$;
3 **for** $t = 1, 2, \cdots,$ **do**
4   **if** $t = mT + s$ *for integers $m \geq 0$ and $1 \leq s < T$* **then**
5          $\lfloor$ Choose $I_t^{(j)}$ randomly according to the mixed strategy $P_m^{(j)}$;
6   **if** $t = mT$ *for integers $m \geq 1$* **then**
7   **if** $\max_{i_j=1,\cdots,K_j} r_{m,i_j} > \rho$ **then**
8              $\lfloor$ Choose $P_m^{(j)}$ randomly according to the uniform distribution over $\{1, \cdots, K_j\}$;
9   **else**
10   With probability $1 - \lambda$, let $P_m^{(j)} = P_{m-1}^{(j)}$, otherwise with probability λ, choose $P_m^{(j)}$ randomly according to the uniform distribution over $\{1, \cdots, K_j\}$;

Then the regret $\hat{r}_{m,i_j}^{(j)}$ is computed as[CL06]

$$\hat{r}_{m,i_j}^{(j)} = \frac{1}{T - K_j n_j} \sum_{s=(m-1)T+1}^{mT} X^{(j)}(I_s) \mathbb{I}_{U_{j,s}=0} - \frac{1}{n_j} \sum_{s=(m-1)T+1}^{mT} X^{(j)}(I_s^-, i_j) \mathbb{I}_{U_{j,s}=i_j}.$$

(4.28)

With proper choices of parameters, the mixed strategy profile will be an approximate Nash equilibrium according to the following theorem:

Theorem 4.11 ([CL06], Theorem 7.8) *For almost all games, there exists a positive number ε_0 and positive constants $c_1, \ldots, c_4$, such that for all $\varepsilon < \varepsilon_0$, if the experimental regret test procedure is run with parameters*

$$\rho \in (\varepsilon, \varepsilon + \varepsilon^{c_1}), \lambda \leq c_2 \varepsilon^{c_3}, T \geq -\frac{1}{2(\rho - \varepsilon)^2} \log(c_4 \varepsilon^{c_3})$$

then for all periods $M \geq \log(\varepsilon/2)/\log(1 - \lambda_c^K)$, $K_c = \sum_{j=1}^N K_j$,

$$P_M(\bar{N}_\varepsilon) = \mathbb{P}[\sigma_{MT} \notin N_\varepsilon] \leq \varepsilon,$$

where N_ε and $\bar{N}_\varepsilon$ denote the set of ε-Nash equilibria and its complement, respectively. Moreover, σ_{MT} denotes the joint action profile of players after period M.

That is, if the parameters of the ERT algorithm are set properly, then the played mixed strategy profile is an approximate *Nash equilibrium* for almost all the time in the long run.

4.4 Summary

In this chapter, we investigated the *adversarial MAB* problem. We first introduced the notions of *external* and *internal* regrets as a measure of the performance of the action selection strategy. We then discussed some well-known algorithms for this problem and provided the asymptotic performance bounds for these strategies. We also generalized the adversarial MAB problem to the multiplayer case. Using the repeated game theory, we showed that some strategies can achieve the correlated or approximate Nash equilibrium for this game.

References

[AB09]   Jean-Yves Audibert and Sébastien Bubeck. "Minimax policies for adversarial and sto-chastic bandits". In: *COLT*. 2009, pp. 217–226.

[AB10]   Jean-Yves Audibert and Sébastien Bubeck. "Regret bounds and minimax policies under partial monitoring". In: *Journal of Machine Learning Research* 11.Oct (2010), pp. 2785–2836.

[Aue+02] Peter Auer et al. "The nonstochastic multiarmed bandit problem". In: *SIAM Journal on Computing* 32.1 (2002), pp. 48–77.

[Aue+95] Peter Auer et al. "Gambling in a rigged casino: The adversarial multiarmed bandit prob-lem". In: *Foundations of Computer Science, 1995. Proceedings., 36th Annual Symposium on*. IEEE. 1995, pp. 322–331.

[CL06]   Nicolo Cesa-Bianchi and Gábor Lugosi. *Prediction, learning, and games*. Cambridge university press, 2006.

[FV99]   Dean P Foster and Rakesh Vohra. "Regret in the on-line decision problem". In: *Games and Economic Behavior* 29.1 (1999), pp. 7–35.

[Han57]   James Hannan. "Approximation to Bayes risk in repeated play". In: *Contributions to the Theory of Games 3* (1957), pp. 97–139.

[Hoe63]   Wassily Hoeffding. "Probability inequalities for sums of bounded random variables". In: *Journal of the American statistical association* 58.301 (1963), pp. 13–30.

[KE07]   Jussi Kujala and Tapio Elomaa. "Following the perturbed leader to gamble at multi-armed bandits". In: *International Conference on Algorithmic Learning Theory*. Springer. 2007, pp. 166–180.

[LW89]   Nick Littlestone and Manfred K Warmuth. "The weighted majority algorithm". In: *Foun-dations of Computer Science, 1989., 30th Annual Symposium on*. IEEE. 1989, pp. 256–261.

[SL05]   Gilles Stoltz and Gábor Lugosi. "Internal regret in on-line portfolio selection". In: *Machine Learning* 59.1-2 (2005), pp. 125–159.

Part II
Applications

Chapter 5
Spectrum Sensing and Access in Cognitive Radio Networks

Abstract With tremendous growth in wireless services, the demand for radio spectrum has significantly increased. However, spectrum resources are scarce and most of them have been already licensed to existing operators. Recent studies have shown that despite claims of spectral scarcity, the actual licensed spectrum remains unoccupied for long periods of time. Thus, cognitive radio (CR) systems have been proposed in order to efficiently exploit these spectral holes, when licensed primary users (PUs) are not present. CRs or secondary users (SUs) are wireless devices that can intelligently monitor and adapt to their environment, hence, they are able to share the spectrum with the licensed PUs, operating when the PUs are idle. In this chapter, we discuss spectrum sensing and access in cognitive radio networks. Depending on the nature of the activities in the network, different categories of MAB problems have been formulated. Solutions and regret bounds are provided.

5.1 Introduction

With the prevalence of wireless technologies, radio spectrum is becoming a scare resource. Traditionally, RF spectrum is managed by regulatory agencies through the assignment of fixed portions of spectrum to individual licensees in the form of renewable licenses. Such a regulatory approach suffers from low spectrum utilization. It has been estimated that spectrum utilization in licensed RF bands, varies from 15% to 85% at different geographic locations at a given time [FCC02]. Dynamic spectrum access (DSA) (Fig. 5.1) was first proposed by the Federal Communications Commission (FCC) in 2003 as means to improve the spectrum utilization efficiency in the United States. There are two types of users in DSA, namely, *primary users* (PU) and *secondary users* (SUs). PUs, typically the licensees, have higher priority or legacy rights on the usage of a specific part of the spectrum. SUs on the other hand, have lower priority and access the spectrum opportunistically. In this chapter, the term DSA is used in the narrow sense as the synonym of opportunistic spectrum access, where SUs are allowed to access the spectrum allocated to the PUs as long as they do not pose excessive interference to the PUs [ZS07]. DSA directly targets

© Springer International Publishing AG 2016
R. Zheng and C. Hua, *Sequential Learning and Decision-Making in Wireless Resource Management*, Wireless Networks, DOI 10.1007/978-3-319-50502-2_5

Fig. 5.1 An Illustration of dynamic spectrum access. Second users access spectrum holes in different space-time-frequency from the primary user

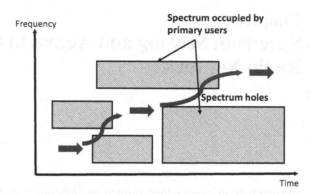

spatial and temporal spectrum white space by allowing secondary users to identify and exploit local and instantaneous spectrum availability in a nonintrusive manner.

The terms of software-defined radio (SDR) and cognitive radio (CR) were independently coined by Mitola [Mit93]. SDR, is generally a multiband radio that supports multiple air interfaces and protocols and is reconfigurable through software run on DSP or general-purpose microprocessors [Mitt00b]. CR, built on a software radio platform, is a context-aware intelligent radio potentially capable of autonomous reconfiguration by learning from and adapting to the communication environment [Mitt00a]. The software programmability and the potential to adapt to its radio environment make CR an ideal candidate technology for DSA.

The overall design objective of DSA is to provide sufficient benefits to secondary users while protecting spectrum licensees from interference. In general, it consists of three components: spectrum opportunity identification, spectrum opportunity exploitation and regulatory policy [ZS07]. In spectrum opportunity identification, a channel is considered an opportunity to a pair of transmitter A and receiver B if they can communicate successfully over this channel while limiting the interference to primary users below a prescribed level determined by the regulatory policy. The interference constraint is typically expressed in terms of the maximum interference power level η perceived by an active primary receiver and the maximum probability ς that the interference level at an active primary receiver may exceed η. Spectrum opportunity identification can be done at an individual node or collaboratively among multiple nodes. It is often accomplished by spectrum sensing using energy detection [TS05] or (cyclostationary) feature detection [CGD07]. Once spectrum opportunity is identified, an SU needs an access strategy to determine whether to transmit over a particular channel. In the presence of multiple SUs, medium access protocols are needed to arbitrate the access to the available channel. The mechanism of spectrum sensing before spectrum access is also known as "Listen-before-Talk" (LBT).

As evident from the discussion, the key challenge in designing efficient DSA schemes lies in the uncertainty in PU activities. To make the matter more complex, the uncertainty is typically location dependent. Thus, each SU may have a different view of the spectrum. When an SU detects the activity in one channel and tries to access it when available, there is an opportunity loss due to possible spectrum opportunities in other channels. In absence of prior knowledge, trade-offs exist in finding the statistics of all channels ("exploration") and in accessing the most available channel known so far ("exploitation"). Fortunately, we have already at our disposal sequential learning tools to address such trade-offs. In particular, SUs are the players in the bandit problem whereas the PU channels are arms. A successful spectrum access allows an SU (or more precisely a pair of SU transceivers) to communicate a certain amount of information over the channel and to accumulate some reward. SUs need to learn the PU activity statistics to maximize the cumulative reward.

In the rest of the chapter, we first present the problem formulation based on slotted spectrum sensing and access mechanism with unknown PU activities. Next, we discuss several sequential learning strategies that are tailored to address different assumptions on the PU processes.

5.2 Problem Formulation

5.2.1 Single SU with IID Rewards

Consider a slotted system of slot length T. In each slot, an SU needs to pick a channel k out of K channels and senses for τ time. If the channel is deemed to be idle (or, transmissions of the SU are unlikely to cause substantial interference to potential PUs), the SU can transmit until the end of the slot. Otherwise, the SU needs to wait till the start of the next slot to repeat the process. In this scheme, τ and T are given as inputs. In practice, the choice of τ depends on a number of factors including the spectrum sensing mechanism used, limitations on the interference to PUs, the distance and transmission power level of potential PUs, etc. T is the result of regulatory policies that mandate SUs to vacate the channel within a certain period of time if PUs become active. The basic slotted spectrum sensing and access scheme is illustrated in Fig. 5.2.

Assume at each time slot, channel k is free with probability θ_k, $k = 1, 2, \ldots, K$. Let $Z_k(t)$ be an IID random variable that equals 1 if channel k is free at time slot t and equals 0 otherwise. Hence, given θ_k, $Z_k(t)$ is a Bernoulli random variable. If $Z_k(t)$ is 1 then an SU can transmit λ bits over this time; otherwise, no information

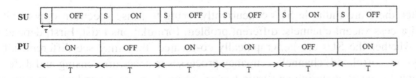

Fig. 5.2 An illustration of Listen-before-talk spectrum sensing and access

can be transmitted. In the single SU case, let I_t be the channel selected at time t. The reward at time t is thus given by $\lambda Z_{I_t}(t)$ and the cumulative rewards over n slots are,

$$W_n = \sum_{t=1}^{n} \lambda Z_{I_t}(t).$$

The goal is to maximize expected reward $\mathbb{E}[W_n]$ over n plays where the exact values of θ_k's are not known. Alternatively, we can define the expected loss or regret as $L_n = \sum_{t=1}^{n} \lambda \theta^* - \mathbb{E}[W_n]$, where $\theta^* = \max_k \theta_k$. The basic problem formulation assumes IID Bernoulli reward processes when a single SU tries to perform spectrum sensing and access. Several important variations of the basic problem have been investigated in the literature.

5.2.2 Single SU with Markov Reward Processes

Besides IID Bernoulli reward processes, Markov reward process is of great interest. Finite-state Markov processes have commonly been employed in wireless communication to characterize variations in channel capacity. One well-known model is the Gilbert–Elliot model, a two-state Markov chain corresponding to the good and bad states [Gil60].

Let us consider the occupancy of channel k modeled by an irreducible, aperiodic, reversible Markov chain on a finite-state Markov chain with parameters $P_k = \left(p_{s,s'}^k : s, s' \in S_k\right)$, where S_k is a finite state space. Let π_k be the stationary distribution of Markov chain P_k and $\mu_k = \sum_{x \in S_k} \pi_x^k$ gives the expected reward.

Two types of Markov reward processes can be considered, namely, rested and restless. In a rested Markov reward process, the chain makes a transition only when the corresponding arm is played. In a restless Markov reward process, the occupancy of channel k evolves according to P_k regardless of the actions of the SU. In the cognitive radio context, since PU activities are assumed to be independent from SU's spectrum access activities, the restless Markov MAB formulation is of more relevance.

5.2.3 Multiple SUs

When there are multiple SUs concurrently trying to sense spectrum opportunities and access vacant channels, different problem formulations exist. First, depending on whether the SUs are geographically co-located, they may see different (albeit possibly correlated) channel occupancy processes in spectrum sensing and different channel capacities during spectrum access. Second, the SUs can be cooperative or noncooperative in spectrum sensing. Third, the SUs can be cooperative or competing in spectrum access. In the centralized cases, a central server collects sensing information from the SUs and/or coordinate the spectrum access decisions among the SUs. What is often interesting in practice is distributed algorithms where SUs sense the spectrum and make observations independently, and follow a decentralized protocol in spectrum access.

In presence of multiple SUs, one optimization objective is to maximize the sum throughput. A secondary objective is to allocate the available spectrum resource fairly among the SUs.

5.3 Solution Approaches

In this section, we discuss solution approaches to the DSA problem and its variations. For a single SU and IID reward processes, one can straightforwardly apply policies for stochastic MAB (e.g., UCB-based and ε-greedy) and achieve logarithmic regrets in time. For a single SU and restless Markov reward processes, we can apply the algorithms discussed in Chap. 3 to achieve logarithmic regrets with respect to a single-action optimal policy or sublinear regrets in comparison to an optimal policy for the underlying MDP with known parameters. Therefore, the focus of the remaining sections will be on multiple SUs for cooperative spectrum access.

5.3.1 Cooperative Spectrum Access

Let $m < K$ be the number of *co-located SUs*. Since the SUs are co-located, they observe identical PU activities and have identical rewards when accessing the same channel as the same time. For ease of discussion, assume that the mean reward of each PU channels are distinctive. Without loss of generality, assume the indices of the PU channels are sorted according to the descending order of the mean rewards, namely, $\mu_1 > \mu_2 > \ldots > \mu_K$. When channel k is idle, if multiple SUs (≥ 2) try to access it at the same time, the instantaneous reward is zero—this is called a *collision*. To maximize the sum throughput, a centralized coordinator should operate SUs on channel $1, 2, \ldots, M$.

In the decentralized setting, SUs need to follow a common protocol to minimize the chance of collisions. The most straightforward approach is to assign fixed ranks to SUs from $1, 2, \ldots, M$ such that an SU with rank l accesses channel l (the lth most available channel). However, this scheme suffers from poor fairness and poor adaptability if the number of SUs changes. To handle the fairness concern, in [LZ10, VLZ13], SUs take turns in accessing the top channels. As an example, consider $m = 2$. The time slots are divided into even and odd slots. In the even slots, SU 1 targets the best channel and SU 2 targets the second best channel. In the odd slots, the opposite happens. Alternatively, without such a pre-agreement, when an SU joins the system, it randomly generates a local offset uniformly drawn from

$\{0, 1, \ldots, m - 1\}$ and plays one round of the top m arms starting from the offset. For example, if the offset is 1, the SU will sense and access channel 2, 3, ..., m, 1 in consecutive slots. If no collision occurs, the SU keeps this offset. Otherwise, in the next round, it randomly choose a new offset. It is easy to see that eventually all SUs will have distinctive offsets and no collisions occur. This scheme allows addition and removal of SUs. However, when m decreases, the resulting scheme is suboptimal as SUs needlessly spend time on worse channels. When m increases, collisions always occur unless m can be updated accordingly. Furthermore, the scheme assumes that all SUs agree upon the ranks of the channels. This is not always feasible in absence of an consensus protocol since the empirical mean estimates of SUs toward the same channel may differ.

An alternative approach is to adopt a distributed algorithm for bipartite matching. Recall that to maximize sum throughput, each SU should access a distinct channel among the top m channels. Here, we model the SUs and the channels as two disjoint subsets of vertices on a fully connected bipartite graph (Fig. 5.3). The m SUs are on one side and the K channels are on the other side. Each SU i has an estimated $\hat{\mu}_{i,j}$, the empirical mean for arm j. Each SU only knows its own estimations. Denote by k^{**} a matching that maximizes $\sum_{i,j} \hat{\mu}_{i,j} x_{i,j}$, where $x_{i,j}$ are binary variables and $\sum_i x_{i,j} \leq 1$ and $\sum_j x_{i,j} \leq 1$. An ε-optimal matching k^* satisfies, $\sum_i \hat{\mu}_{i,k^{**}(i)} - \sum_i \hat{\mu}_{i,k^*(i)} \leq \varepsilon, 1 \leq i \leq m, 1 \leq j \leq K$. Bertsekas' auction algorithm [Ber92] can find the ε-optimal matching on a weighted bipartite graph and is amiable to distributed implementation. The pseudocode of the algorithm is given in 5.1. In the description, $2max_j$ stands for the second highest maximum over all j's.

Bertsekas' ε-optimal auction algorithm results in a stable matching, i.e., no SU has any incentive to deviate from its selection. One key advantage is that the SUs do not exchange its (private) reward information. It is applicable to scenarios where the SUs have different rewards (e.g., throughputs) from the same channel. However, the

Fig. 5.3 Spectrum access as a bipartite marching problem. The SUs and PU channels two disjoint subsets connected by a fully connected bipartite graph. The edge weight between SU i and PU channel j corresponds to the empirical mean of the rewards SU i observes on channel j. The maximum weighted matching is highlighted in bold

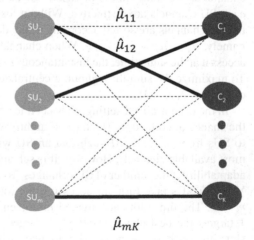

Algorithm 5.1: Bertsekas' ε-optimal Auction Algorithm

---

1 **Init:** *All SU i set prices* $p_j = 0, j = 1, 2, \ldots K$
2 **while** *prices change* **do**
3   SU i communicates his preferred arm j_i^* and bid
        $b_i = \max_j(\hat{\mu}_{ij} - p_j) - 2max_j(\hat{\mu}_{ij} - p_j) + \frac{\varepsilon}{M}$ to all other SUs;
4   Each player determines on his own if he is the winner i_j^* on arm j with the highest bid;
5   All players set prices $p_j = \hat{\mu}_{i_j^*,j}$;

---

algorithm does not consider fairness across multiple SUs. When m or $\hat{\mu}$ changes, the iterative bidding process needs to be executed again.

5.3.2 Distributed Learning and Allocation

Now we are in the position to discuss algorithms for decentralized spectrum sensing and access without prior knowledge of the reward processes. All algorithms discussed in this section assume synchronized slots among all SUs.

In [KNJ12], Kalathil et al. propose a scheme based on UCB for IID reward processes. Each SU first senses each of the K channels once. In a slot, an SU j tries to sense and access a channel with the top l_jth upper confidence bound defined as

$$g_{i,j}(t) := \hat{\mu}_{i,j}(t) + \sqrt{\frac{2 \log t}{T_{i,j}(t)}}, \qquad (5.1)$$

where $\hat{\mu}_{i,j}(t)$ is the average reward for SU j on channel i and $T_{i,j}(t)$ is the number of times SU j senses channel i till time t. There are three possible outcomes. First, the SU experiences a collision. It will randomly choose a new l_j. Second, the SU observes a busy channel. Third, the SU observes an idle channel and transmits with no collision. In the later two cases, the SU updates the g statistics for all channels accordingly. It has been shown that the policy incurs logarithmic regrets in time.

In [TL12], Tekin and Liu extended the regenerative cycle algorithm (Sect. 3.2.2.1) to the decentralized multiuser dynamic spectrum access scenario for restless Markov reward processes. An SU selects a channel among its perceived top m channels based on the statistics in (5.1). If collisions occur, it randomly chooses a channel of a different rank among the top m channels. Otherwise, it stays with the channel of the same rank.

The time-division fair sharing (TDFS)-based policy proposed by Liu and Zhao in [LLZ13] considers the fairness among SUs in the decentralized setting with IID reward processes. The time sequence is divided into m subsequences, in which each player targets at the m best arms in a round-robin fashion with a different offset. The offsets are determined using the procedure described in Sect. 5.3.1. For the ease of discussion, assume all offsets are distinct. Suppose that SU 1 has offset 0, i.e., it targets at the kth ($1 \leq k \leq M$) best arm in the kth sequence. The kth subsequence is then divided into C_{k-1}^K mini sequences, each associated with a subset of $K - k + 1$ channels after removing the top $k - 1$ channels with higher ranks (ranks are established with a similar statistics like UCB). In each of the subsequences, a single-player policy for the stochastic MAB problem is applied. It has been proven that the TDFS-based policy can achieve a logarithmic regret bound.

Using the decentralized bipartite matching algorithm such as Bertsekas' ε-optimal auction algorithm as an inner procedure, Kalathil et al. devised distributed policies that can be applied to IID reward and Markov reward processes in [KNJ12].

5.4 Summary

In this chapter, we presented the formulation of spectrum access and spectrum sensing in cognitive radio networks. Different policies have been discussed that target different types of primary user processes. Recently, there have been renewed interests in applying the CR technology to improve spectrum utilization. These are in part fused by the advancement of wireless physical layer techniques such as massive multiple-input and multiple-output (MIMO), millimeter-wave, full-duplex communication, which warrants reexamination of basic assumptions such as Listen-before-talk and isotropic SU-to-PU inference patterns [Wu+13, HZD16].

References

[Ber92]   Dimitri P Bertsekas. "Auction algorithms for network flow problems: A tutorial intro-
           duction". In: Computational optimization and applications 1.1 (1992), pp. 7–66.
[CGD07]   Hou-Shin Chen, Wen Gao, and David G Daut. "Spectrum sensing using cyclostationary
           properties and application to IEEE 802.22 WRAN". In: IEEE GLOBECOM 2007-
           IEEE Global Telecommunications Conference. IEEE. 2007, pp. 3133–3138.
[FCC02]   FCC Spectrum Policy Task Force. Report of the spectrum efficiency working group.
           http://transition.fcc.gov/sptf/reports.html. 2002.
[Gil60]   Edgar N Gilbert. "Capacity of a Burst-Noise Channel". In: Bell system technical journal
           39.5 (1960), pp. 1253–1265.
[HZD16]   Mohamed Hammouda, Rong Zheng, and Timothy N. Davidson. "Fullduplex spectrum
           sensing and access in cognitive radio networks with unknown primary user activi-
           ties". In: 2016 IEEE International Conference on Communications, ICC 2016, Kuala
           Lumpur, Malaysia, May 22-27, 2016. 2016, pp. 1–6.
[KNJ12]   D. Kalathil, N. Nayyar, and R. Jain. "Decentralized learning for multiplayer multi-
           armed bandits". In: Decision and Control (CDC), 2012 IEEE 51st Annual Conference
           on. 2012, pp. 3960–3965.
[LLZ13]   Haoyang Liu, Keqin Liu, and Qing Zhao. "Learning in a changing world: Restless mul-
           tiarmed bandit with unknown dynamics". In: Information Theory, IEEE Transactions
           on 59.3 (2013), pp. 1902–1916.

[LZ10]   K. Liu and Q. Zhao. "Distributed Learning in Multi-Armed Bandit With Multiple
            Players". In: IEEE Transactions on Signal Processing 58.11 (Nov. 2010), pp. 5667–
            5681. ISSN: 1053-587X.
[Mitt00a]   Joseph Mitola III. "Cognitive radio". PhD thesis. Royal Institute of Technology, 2000.
[Mitt00b]   J Mitola. Software Radio: Wireless Architecture for the 21st Century. Chichester, UK:
            John Wiley & Sons, Inc, 2000.
[Mit93]   J. Mitola. "Software radios: Survey, critical evaluation and future directions". In: IEEE
            Aerospace and Electronic Systems Magazine 8.4 (Apr. 1993), pp. 25–36.
[TL12]   Cem Tekin and Mingyan Liu. "Online learning of rested and restless bandits". In: IEEE
            Transactions on Information Theory 58.8 (2012), pp. 5588–5611.
[TS05]   Rahul Tandra and Anant Sahai. "Fundamental limits on detection in low SNR under
            noise uncertainty". In: 2005 International Conference on Wireless Networks, Commu-
            nications and Mobile Computing. Vol. 1. IEEE. 2005, pp. 464–469.
[VLZ13]   Sattar Vakili, Keqin Liu, and Qing Zhao. "Deterministic sequencing of exploration and
            exploitation for multi-armed bandit problems". In: IEEE Journal of Selected Topics in
            Signal Processing 7.5 (2013),pp. 759–767.
[Wu+13]   Qihui Wu et al. "Spatial-temporal opportunity detection for spectrumheterogeneous
            cognitive radio networks: two-dimensional sensing". In: IEEE Transactions on Wireless
            Communications 12.2 (2013), pp. 516–526.
[ZS07]   Qing Zhao and Brian M Sadler. "A survey of dynamic spectrum access". In: IEEE
            signal processing magazine 24.3 (2007), pp. 79–89.

Chapter 6
Sniffer-Channel Assignment in Multichannel Wireless Networks

Abstract Passive monitoring is a technique where a dedicated set of hardware devices called sniffers, are used to monitor activities in wireless networks. Distributed sniffers cooperatively monitor PHY and MAC characteristics, and interactions at various layers of the protocol stacks, both within a managed network and across multiple administrative domains. One important issue in designing a monitoring network is to determine which set of frequency bands each sniffer should operate on to maximize the total amount of information gathered. In this chapter, we discuss the sniffer-channel assignment problem in multichannel wireless networks and its formulation as a stochastic MAB problem. Compared to the basic stochastic MAB problem, the key challenge lies in the consideration of correlations among multiple sniffers' observations. We present efficient learning algorithms and their regret bounds. Important issues that often arise from practical deployments such as switching overhead and computation efficiency are also considered.

6.1 Introduction

Deployment and management of wireless infrastructure networks (WiFi, WiMax, wireless mesh networks) are often hampered by the poor visibility of PHY and MAC characteristics, and complex interactions at various layers of the protocol stacks both within a managed network and across multiple administrative domains. In addition, today's wireless usage spans a diverse set of QoS requirements from best-effort data services, to VOIP and streaming applications, making the task of managing the wireless infrastructure even more difficult, due to the additional constraints posed by QoS sensitive services. Monitoring the detailed characteristics of an operational wireless network is critical to many system administrative tasks including, e.g., fault diagnosis, resource management, and critical path analysis for infrastructure upgrades.

Passive monitoring is a technique where a dedicated set of hardware devices called *sniffers*, or monitors, are used to monitor activities in wireless networks. These devices capture transmissions of wireless devices or activities of interference sources in their vicinity, and store the information in trace files, which can be analyzed distributively or at a central location. Wireless monitoring [YYA04, Yeo+05, Rod+05,

© Springer International Publishing AG 2016

R. Zheng and C. Hua, *Sequential Learning and Decision-Making in Wireless Resource Management*, Wireless Networks, DOI 10.1007/978-3-319-50502-2_6

Che+06, CZ09] has been shown to complement wire side monitoring using SNMP and basestation logs since it reveals detailed PHY (e.g., signal strength, spectrum density) and MAC behaviors (e.g., collision, retransmissions), as well as timing information (e.g., backoff time), which are essential for wireless diagnosis. The architecture of a canonical monitoring system consists of three components, (1) sniffer hardware, (2) sniffer coordination and data collection, and (3) data processing and mining.

Depending on the type of networks being monitored and hardware capability, sniffers may have access to different levels of information. For instance, spectrum analyzers can provide detailed time- and frequency- domain information. However, due to the limit of bandwidth or lack of hardware/software support, they may not be able to decode the captured signal to obtain frame level information on the fly. Commercial-off-the-shelf network interfaces such as WiFi cards on the other hand, can only provide frame level information.[1] The volume of raw traces in both cases tends to be quite large. For example, in the study of our campus WLAN, 4 million MAC frames have been collected per sniffer per channel over an 80-minute-period resulting in a total of 8 million distinct frames from four sniffers. Furthermore, due to the propagation characteristics of wireless signals, a single sniffer can only observe activities within its vicinity. Observations of sniffers within close proximity over the same frequency band tend to be highly correlated. Therefore, two pertinent issues need to be addressed in the design of passive monitoring systems; (1) what to monitor, and (2) how to coordinate the sniffers to maximize the amount of captured information.

We assume a generic architecture of passive monitoring systems for wireless infrastructure networks, which operate over a set of contiguous or noncontiguous channels or bands.[2] An important issue is to determine which set of frequency bands each sniffer should operate on to maximize the total amount of information gathered. This is called the *sniffer-channel assignment* problem or *channel assignment* problem for short. It is a challenging problem for a number of reasons. First, monitoring resources are limited, and thus it is infeasible to monitor all channels at all locations at all times. Second, intelligent channel assignment requires the knowledge of usage patterns, i.e., the likelihood of the occurrence of interesting events. These are not known a *priori*. An interesting trade-off arises between assigning sniffers to channels known to be the busiest based on current knowledge, versus exploring channels that are under-sampled. Third, in practical systems, channel switching is not instantaneous. For example, [SB08] reports on the 802.11b/g AR5212 chipset that the channel switching operation requires a full hardware reset that incurs a delay of approximately 1.2 ms. An additional delay of about 3.2 ms is introduced in the channel switching operation due to other system operations, such as flushing any

---

[1]Certain chip sets and device drivers allow inclusion of header fields to store a few physical layer parameters in the MAC frames. However, such implementations are generally vendor and driver dependent.

[2]A channel can be a single frequency band, a code in CDMA systems, or a hopping sequence in frequency hopping systems.

pending transmission buffers in the hardware queues and waiting for any pending transmit or receive direct memory access (DMA) operations to finish. Measurements on USRP2™ [LLC], a software defined radio platform indicate a latency on the order of hundreds of milliseconds when shifting the central frequency of the spectrum analyzer implemented in GNURadio [GNU]. During channel switching, packet receptions and transmissions are not possible. As a result, frequent switching is undesirable. Fourth, the sniffer-channel assignment decisions need to be made in timely manner. Optimal decisions that require the solution to NP-hard problems do not scale well as the network size grows.

To determine the optimal allocation of monitoring resources to maximize captured information, Shin and Bagchi consider the selection of monitoring nodes and their associated channels for monitoring wireless mesh networks [SB09]. The optimal monitoring is formulated as a maximum coverage problem with group budget constraints, which was previously studied by Chekuri and Kumar in [CK04]. In [Chh+10], we introduced a quality of monitoring (QoM) metric defined by the expected number of active users monitored, and investigated the problem of maximizing QoM by judiciously assigning sniffers to channels based on knowledge of user activities in a multichannel wireless network. We assume the *sniffer-centric model* that utilizes binary channel information only(active or not) at a sniffer.

The above works assume that certain statistics regarding the users' activity are given [SB09, CK04] or can be inferred [Chh+10]. When such statistics are not known a priori, sequential learning is needed. Sequential decision making in presence of uncertainty, faces the fundamental trade-off between *exploration* and *exploitation*. On one hand, it is desirable to put sniffers to the channels where most activities have been observed and thus more information is likely to be gathered (exploitation). On the other hand, exploring the channels that are under-sampled helps to reduce uncertainty and thus avoids being misled by imprecise information. Such trade-offs are fully captured by the MAB problem.

6.2 Problem Formulation

Consider m sniffers monitoring user activities in K channels. A user u can operate in one of K channels, $c(u) \in \mathcal{K} = \{1, \ldots, K\}$. Let p_u denote the transmission probability of user u. We represent the relationship between users and sniffers using an undirected bipartite graph $G = (S, U, E)$, where $S = \{1, \ldots, m\}$ is the set of sniffer nodes and U is the set of users. An edge $e = (s, u)$ exists between sniffer $s \in S$ and user $u \in U$ if u is within the reception range of sniffer s. If transmissions from a user cannot be captured by any sniffer, the user is excluded from G. For every vertex $v \in S \cup U$, we let $N(v)$ denote vertex v's neighbors in G. For users, their neighbors are sniffers, and vice versa. We assume that one sniffer can observe one user at a time. This is consistent with many existing multiple access mechanisms including FDMA and TDMA.

At any point in time, a sniffer can only observe transmissions over a single channel. We will consider *channel assignments* of sniffers to channels, $\mathbf{k} = (k_1, \ldots, k_m)$, where $1 \leq k_i \leq K$. Let $\mathbb{K} = \{\mathbf{k} \mid \mathbf{k} : S \rightarrow \{1, .., K\}^m\}$ be the set of all possible assignments. The set of users a sniffer s can observe is given by $N(s) \bigcap \{u : c(u) = k_s\}$.

6.2.1 Optimal Channel Assignment in the Nominal Form

We first consider the formulation of the optimal sniffer-channel assignment where the graph G and the user activity probabilities $(p_u; u \in U)$ are both known. Since optimal channel assignment with uncertainty is inherently harder than that without uncertainty, determining the complexity of the later provides a baseline understanding of the computational aspect of the former problem.

The objective of optimal channel assignment is to maximize the expected number of active users monitored. Let MAX- EFFORT- COVER (MEC) denote the problem of determining the optimal channel assignment of sniffers to maximize the total weight of users monitored, under the constraint that each sniffer can monitor one of a set of K channels. Note that in MEC, the weights can in fact be any nonnegative values and are not limited to [0, 1]. The MEC problem can be cast as the following integer program (IP).

$$
\begin{aligned}
\max \ & \textstyle\sum_{u \in U} p_u y_u \\
\text{s.t. } & \textstyle\sum_{k=1}^{K} z_{s,k} \leq 1 & \forall s \in S \\
& y_u \leq \textstyle\sum_{s \in N(u)} z_{s,c(u)} & \forall u \in U \\
& y_u, z_{s,k} \in \{0, 1\} & \forall u, s, k.
\end{aligned}
\tag{6.1}
$$

Each sniffer is associated with a set of binary decision variables: $z_{s,k} = 1$ if the sniffer is assigned to channel k; 0, otherwise. Further, y_u is a binary variable (but not a decision variable) indicating whether or not user u is monitored, and p_u is the weight associated with user u. The following result has been proven in [Chh+10]:

Theorem 6.1 (NP-hardness of the maximum-effort-cover problem) *The* MEC *problem is NP-hard with respect to the number of sniffers, even for $K = 2$.*

In other words, the computational complexity for a genie to make the optimal choice with the knowledge of all users' activity grows faster than any polynomial with respect to the number of sniffers, unless $P = NP$. However, when the graph G has some specific structure, there may exist efficient algorithms. For example, when G is restricted to be a complete bipartite graph, it can be shown that MEC reduces to the maximum matching in a transformed bipartite graph, which can be solved in polynomial time.

Approximation algorithms with constant approximation ratios have been devised in [Chh+10]. In particular, a simple greedy algorithm (GREEDY) that incrementally picks an unassigned sniffer and chooses the channel that maximizes the average number of remaining users it monitors is shown to be $\frac{1}{2}$-approximate.

When the graph G and the user activity probabilities are given, the optimal sniffer-channel assignment is fixed. However, when the user activity probabilities are unknown, a learning strategy must try different assignments. Furthermore, intelligent schemes need to be designed to take switching costs into account.

6.2.2 Linear Bandit for Optimal Channel Assignment with Uncertainty

Now, we turn to the optimal channel assignment when there is uncertainty in both G and p_u's. We first define the structure of the instantaneous feedback and payoff of each sniffer.

Let $U_{ik}(t)$ be a nonnegative, integer-valued random variable that denotes the index of the user whose activity sniffer i observes in channel k at time t, or which takes the value of zero if there is no user activity in channel k. For simplicity, we assume that $U(t) = (U_{ik}(t); 1 \leq i \leq m, 1 \leq k \leq K)$ is a sequence of IID random variables. The instantaneous feedback (observations) received under the joint action $\mathbf{k} = (k_1, \ldots, k_m)$ is $Y^\circ_{(k_1,\ldots,k_m)}(t) = (U_{1,k_1}(t), U_{2,k_2}(t), \ldots, U_{p,k_m}(t))$. Let $\mathbb{I}_{\{x\}}$ be the indicator function, where $\mathbb{I}_{\{x\}} = 1$ if x is true; and $\mathbb{I}_{\{x\}} = 0$, otherwise. Note that the indicator $\mathbb{I}_{\{U_{i_1,k_{i_1}}(t)=U_{i_2,k_{i_2}}(t)=\ldots=U_{i_s,k_{i_s}}(t)>0\}}$ is a function of $Y^\circ_{(k_1,\ldots,k_m)}(t)$ and hence can be taken as part of the observation $Y_{(k_1,\ldots,k_m)}(t)$, defined as the collection

$$\left[\mathbb{I}_{\{U_{i_1,k_{i_1}}(t)=U_{i_2,k_{i_2}}(t)=\ldots=U_{i_s,k_{i_s}}(t)>0\}}; \right.$$
$$\left. 1 \leq s \leq m, \ 1 \leq i_1 < \ldots < i_s \leq m \right].$$

We view $Y_{(k_1,\ldots,k_m)}$ as a vector of 2^m binary variables indicating whether the respective collection of sniffers observe the same user. Clearly, there exists a bijection between $Y^\circ_{(k_1,\ldots,k_m)}$ and $Y_{(k_1,\ldots,k_m)}$ (by possibly renaming the users) under the condition that each sniffer can only observe one user at a time.

Note that spatial multiplexing is allowed such that multiple users can be active at the same time in one channel (as long as they are sufficiently far apart geographically). As in Sect. 6.2.1, the payoff upon selecting the joint action is the number of distinct users observed. That is, the joint payoff for selecting channels $\mathbf{k}(t)$ is given by,

$$X_{\mathbf{k}}(t) = |\{U_{1,k_1}(t), \ldots, U_{p,k_m}(t)\}|$$
$$- \mathbb{I}_{\{U_{1,k_1}(t)=0,\ldots,U_{p,k_m}(t)=0\}}$$
$$= \sum_{i=1}^{p} \mathbb{I}_{\{U_{1,k_i}(t)>0\}}$$
$$- \sum_{i,j=1}^{p} \mathbb{I}_{\{U_{i,k_i}(t)=U_{j,k_j}(t)>0\}} \mathbb{I}_{\{k_i=k_j, i \neq j\}} \qquad (6.2)$$
$$\ldots$$
$$- (-1)^P \mathbb{I}_{\{U_{1,k_1}(t)=U_{2,k_2}(t)=\ldots=U_{m,k_m}(t)>0\}}$$
$$\times \mathbb{I}_{\{k_1=k_2=\ldots=k_m\}}.$$

The expected payoff is given by,

$$
\begin{aligned}
&\mathbb{E}\left[X_{\mathbf{k}}(t)\right]\\
&= \sum_{i=1}^{p} \mathbb{P}\left(U_{1,k_i}(t) > 1\right)\\
&\quad - \sum_{i,j=1}^{p} \mathbb{P}\left(U_{i,k_i}(t) = U_{j,k_j}(t) > 0\right) \mathbb{I}_{\{k_i = k_j, i \neq j\}}\\
&\quad \cdots\\
&\quad - (-1)^p \mathbb{P}\left(U_{1,k_1}(t) = \ldots = U_{p,k_m}(t) > 0\right)\\
&\quad \times \mathbb{I}_{\{k_1 = k_2 = \ldots = k_m\}}
\end{aligned}
\tag{6.3}
$$

Define a vector θ, whose components are initially unknown to the learning algorithm, with the following entries:

$$
\begin{aligned}
\mathbb{P}\left(U_{i,k} > 0\right), &\quad 1 \le i \le m, 1 \le k \le K,\\
\mathbb{P}\left(U_{i_1,k} = U_{i_2,k} > 0\right), &\quad 1 \le i_1 < i_2 \le m, 1 \le k \le K,\\
&\vdots\\
\mathbb{P}\left(U_{1,k} = U_{2,k} = \ldots = U_{p,k} > 0\right), &\quad 1 \le k \le K.
\end{aligned}
\tag{6.4}
$$

We introduce the "arm features", $\phi_{\mathbf{k}} \in \mathbb{R}^M$ shown in (6.5), where $M = K(2^m - 1)$. Note that the jth arm feature $\phi_{\mathbf{k},j}$ is uniquely determined by the arm $\mathbf{k} = (k_1, k_2, \ldots, k_m)$. Let $\mathscr{M}_{\mathbf{k}} = \left\{ i : 1 \le i \le M, \phi_{\mathbf{k},i} \neq 0 \right\}$ be the set of nonzero components of feature vector $\phi_{\mathbf{k}}$ and let $M_{\mathbf{k}} = |\mathscr{M}_{\mathbf{k}}|$.

$$
\phi_{\mathbf{k},i} =
\begin{cases}
\mathbb{I}_{\{k_1 = i\}}, & \text{if } 1 \le i \le K;\\
\cdots & \\
\mathbb{I}_{\{k_2 = i - l \cdot K\}}, & \text{if } l \cdot K + 1 \le i \le (l+1) \cdot K;\\
\cdots & \\
-\mathbb{I}_{\{k_1 = k_2 = i - p \cdot K\}}, & \text{if } p \cdot K + 1 \le i \le (p+1) \cdot K;\\
\cdots & \\
-(-1)^p \mathbb{I}_{\{k_1 = k_2 = \ldots = k_m = i - K(2^m - 2)\}}, & \text{if } K(2^m - 2) + 1 \le i \le K(2^m - 1)
\end{cases}
\tag{6.5}
$$

To this end, we can rewrite the expected payoff in MEC as a linear function of the arm feature $\phi_{\mathbf{k}}$,

$$
\mathbb{E}\left[X_{\mathbf{k}}(t)\right] = \theta^T \phi_{\mathbf{k}},
\tag{6.6}
$$

where $(\cdot)^T$ denotes transposition.

Knowing θ suffices to play optimally: An arm with the maximum payoff is given by $\mathbf{k}^* = \arg\max_{\mathbf{k} \in \mathbb{K}} \theta^T \phi_{\mathbf{k}}$ (here, and in what follows, for the sake of simplicity, we assume that there exists a unique optimal arm). Note that this optimization problem is just a trivial reformulation of the MEC problem in Sect. 6.2.1. A reasonable way to estimate the parameter vector θ is to keep a running average for the components of θ. If at time t the agent chose $\mathbf{k}(t) \in \mathbb{K}$ then the current estimate, $\hat{\theta}(t-1)$, can be updated by

$$\hat{\theta}_i(t) = \hat{\theta}_i(t-1) + \frac{1}{N_i(t)} \left(Y_i(t) - \hat{\theta}_i(t-1) \right) \mathbb{I}_{\{i \in \mathcal{M}_{\mathbf{k}(t)}\}},$$

$$N_i(t) = N_i(t-1) + \mathbb{I}_{\{i \in \mathcal{M}_{\mathbf{k}(t)}\}}.$$

(6.7)

Here $N_i(0) = 0$, $\hat{\theta}_i(0) = 0$. Thus, $N_i(t)$ counts the number of times that component i has been observed up to time t. $Y_i(t)$ is defined in (6.2), the binary observation for component i at time t.

Example 4 (Co-located sniffers) When the sniffers are "co-located" or are deployed at close proximity, their observations are identical. Therefore, $U(t)$ will be such that if $k_i = k_j$ then $U_{i,k_i}(t) = U_{j,k_j}(t).$[3] It can be proved that it is strictly better to put different sniffers to different channels, namely, $k_i \neq k_j$, $1 \leq i < j \leq m$. In this case it suffices to estimate $P(U_{ik} > 0)$, i.e., a total of $K \cdot p$ parameters. The problem then becomes essentially the multi-armed bandit problem with multiple plays considered in a number of previous works [AVW87, LZ10, AMT10].

Example 5 (Independent sniffers) The opposite case is when $U_{i,k_i}(t) \neq U_{j,k_j}(t)$ whenever $i \neq j$ and when one of $U_{i,k_i}(t)$ and $U_{j,k_j}(t)$ is nonzero. This happens when all sniffers are guaranteed to observe distinct users (e.g., they are far away from one another). Then, $\mathbb{I}_{\{U_{i_1,k_{i_1}} = U_{i_2,k_{i_2}} = \ldots = U_{i_s,k_{i_s}} > 0\}} = 0$, $2 \leq s \leq m$, $1 \leq i_1 <, \ldots, < i_s \leq m$. Therefore, the number of parameters is reduced to $K \cdot p$ and each sniffer can decide independently which channel to monitor. Thus the, problem reduces to p independent K-arm bandit problems.

In practice, sniffers are deployed distributively. Their observations are typically correlated but nonidentical. This motives us to consider the optimal channel assignment in general configurations. The learning efficiency of a policy is evaluated in terms of its pseudo-regret given by

$$R_n^\pi = \mathbb{E}\left[\sum_{t=1}^n \left\{ \max_{\mathbf{k} \in \mathcal{A}} \phi_{\mathbf{k}}^T \theta - \phi_{\mathbf{k}(t)}^T \theta \right\} \right],$$

(6.8)

where $\mathbf{k}(t)$ is the assignment selected at time t. The goal of an optimal monitoring policy π is to determine a sequence of actions in $\mathbb{K}$ over time such that R_n^π is minimized.

6.2.3 Extensions

In this section, we discuss some variants of the base problem motivated by practical constraints faced by operational wireless systems.

---

[3]Clock synchronization among sniffers can be achieved online or offline using methods such as in [EGE02].

6.2.3.1   Consideration of Switching Regrets

To define the switching regret, let

$$S_n(\mathbf{k}) = \sum_{t=1}^{n} \mathbb{I}_{\{\mathbf{k}(t)=\mathbf{k}, \mathbf{k}(t+1)\neq\mathbf{k}\}}$$

count the number of switches from the joint action $\mathbf{k}$ to some other action during the first n rounds. The switching regret is

$$SW_n^{\pi} = C_{sw} \sum_{\mathbf{k}\in\mathscr{A}} \mathbb{E}\left[S_n(\mathbf{k})\right], \tag{6.9}$$

where C_{sw} is the switching cost. Note we assume that the switching cost is constant across all pairs of joint actions. This is reasonable in a synchronous system where all sniffers coordinate the onsets of monitoring.

An optimal monitoring policy π determines a sequence of actions in $\mathbb{K}$ over time such that the expected *total* regret is minimized:

$$Q_n^{\pi} = R_n^{\pi} + SW_n^{\pi}. \tag{6.10}$$

6.2.3.2   Approximate Online Learning

In devising the optimal policies to minimize the pseudo-regrets, in each round, typically a subproblem of finding the best arm according to some index statistics needs to be solved. Such an optimization problem may be NP-hard in itself. It is therefore useful to consider approximation policies that are more computational efficient. Clearly, the regret incurred by an approximation policy is likely to be linear in time. Thus, a more meaningful objective is to minimize the growth of regret with respect to an offline approximation algorithm.

Denote k^g the assignment (or arm) chosen by an approximation algorithm g with complete information. We define the regret of policy π relative to policy g as

$$R_g^{\pi}(n) = \mathbb{E}\left[\sum_{t=1}^{n} \left\{\phi_{\mathbf{k}^g}^T\theta - \phi_{\mathbf{k}(t)}^T\theta\right\}\right]. \tag{6.11}$$

The objective is to find a policy that incurs sublinear regrets in time.

6.3 Solution Approaches

One challenge in applying known index-based policies such as the UCB and ε-greedy algorithms in Sect. 2.3 to the sniffer-channel assignment lies in the sheer number of arms involved. For m sniffers operating in K channels, the size of the action space K^m. Even playing each arm once, a total of K^m round is needed. The linear bandit formulation in (6.6) represents the reward of an arm as a product of its arm feature and the unknown vector θ of size $K(2^m - 1)$. The size of parameter space to be learned has been greatly reduced but remains to be exponential with respect to the number of sniffers. In [ASZ11], Arora et al. introduced the notation of spanner arms to allow effective exploration and proposed two index-based policies to the linear bandit problem.

6.3.1 Spanners

Since some arms reveal information about other arms, it might be possible to identify a restricted set $\zeta \subset \mathbb{K}$, which is much smaller than $\mathbb{K}$, so that playing only arms in ζ gives sufficient information to identify the optimal arm. A sufficient condition for this is that $\cup_{k \in \zeta} \mathcal{M}_k = \{1, \ldots, M\}$. This condition ensures that by choosing an appropriate arm in ζ, any component of $X(t)$ can be observed, which is clearly sufficient to identify θ. Since exploration is generally costly, the set ζ is ideally chosen to be small. In the monitoring problem ζ can be chosen to be $\zeta = \{(k, \ldots, k) : 1 \leq k \leq K\}$, i.e. all the sniffers assigned to the same channel to cover $(2^m - 1)$ parameters, whose cardinality is $K \ll K^m = |\mathbb{K}|$. The set ζ is called a *spanning set* or a *spanner* and its elements are called *spanner arms*.

6.3.2 An Upper Confidence Bound (UCB)-Based Policy

The first policy, called *UCB-Spanner*, is similar to UCB1 in Sect. 2.3.1. The main difference is that in the initialization stage, we only play each of the spanners once. Formally, the algorithm first plays each arm in ζ once and then at time $t \geq |\zeta| + 1$ chooses

$$\mathbf{k}(t) = \arg \max_{k \in \zeta} V_k(t - 1),$$

where

$$V_k(t - 1) = \hat{\mu}_k(t - 1) + \sum_{i \in \mathcal{M}_k} \sqrt{\frac{\rho \ln t}{N_i(t - 1)}},$$

$$\hat{\mu}_k(t - 1) = \hat{\theta}(t - 1)^\top \phi_k.$$

After playing $\mathbf{k}(t)$ and observing $(Y_i(t); i \in \mathcal{M}_{\mathbf{k}(t)})$ the parameter estimate is updated using (6.7). Then, the process is repeated. Theorem 6.2 gives the regret bound of this algorithm [ASZ11]. From (6.12), we see that the regret grows logarithmically in time but exponentially in the number of sniffers since $M = K(2^m - 1)$.

Theorem 6.2 (Regret bound of UCB-Spanner) *Choose any ρ that satisfies $\rho > 1/1.99$. Then, there exists a constant $C > 0$ (which may depend on ρ) such that for all $n \geq 1$, the expected regret of UCB-Spanner satisfies*

$$R_n^{\text{UCB-Spanner}} \leq 4M \Delta_{\max} \left(\max_{\mathbf{k}:\Delta_{\mathbf{k}}>0} \frac{M_{\mathbf{k}}}{\Delta_{\mathbf{k}}} \right)^2 \rho \ln n + C, \tag{6.12}$$

where $\Delta_{\max} = \max_{\mathbf{k}} \Delta_{\mathbf{k}}$.

6.3.3 An ε-Greedy Algorithm with Spanner

The policy considered here is a variant of ε-greedy in Sect. 2.3. The standard ε-greedy algorithm for bandit problems chooses with probability ε uniformly at random some arm (i.e., it "explores" with probability ε) and it chooses the arm with the highest estimated payoff otherwise. When ε is appropriately scheduled (basically, one needs $\varepsilon = \varepsilon_n = c/n$ with an appropriately selected constant $c > 0$) this policy can also achieve a logarithmically bounded expected regret just like UCB1 [ABF02].

Since in the sniffer-channel problem, the arms are correlated and when an arm is chosen one receives some additional information in addition to the payoffs, one may restrict the set of arms explored to a spanner ζ. We expect that performance will improve if $|\zeta| \ll |\mathbb{K}|$ since then one "pays less" for the exploration steps.

Formally, the algorithm works as follows: Choose a spanner $\zeta \subset \mathbb{K}$ and a sequence $(\varepsilon_t; t \geq 1)$, $\varepsilon_t \in [0, 1]$. In the initialization phase explore each arm in ζ once and initialize the parameter estimates $\hat{\theta}(\cdot)$ based on the information received. After the exploration phase, at time $t \geq |\zeta|$, the arm to be played is decided by first drawing a random number U_t from the uniform distribution over $[0, 1]$. If $U_t \leq \varepsilon_t$ then $\mathbf{k}(t)$ is chosen uniformly at random from ζ. Otherwise, $\mathbf{k}(t) = \arg\max_{\mathbf{k}} \hat{\mu}_{\mathbf{k}}(t - 1)$, where $\hat{\mu}_{\mathbf{k}}(t - 1) = \hat{\theta}(t - 1)^\top \phi_{\mathbf{k}}$. After playing $\mathbf{k}(t)$ and observing the feedback, the parameters are updated using (6.7).

The next theorem gives a bound on the regret of this policy:

Theorem 6.3 (Regret bound of ε-greedy with Spanner) *Let*

$$\varepsilon_n = \min \left\{ 1, \frac{c}{n} \right\}, \ n > |\zeta|, \tag{6.13}$$

where $c > 0$ is a tuning parameter. Then, assuming that $c > \min(10|\zeta|, \frac{4|\zeta|}{d^2})$, where $d = \min_{\mathbf{k}:\Delta_{\mathbf{k}}>0} \Delta_{\mathbf{k}}$, the expected regret of ε-greedy satisfies

$$\mathbb{E}\left[R_n^{\varepsilon-\text{greedy}}\right] \le c\ln(n+1) + O(1). \tag{6.14}$$

From the point of view of minimizing the leading term, the best choice of c is $\min(10|\zeta|, \frac{4|\zeta|}{d^2})$. With such a choice, we see that the leading term of regret scales linearly with $|\zeta|$, and not with $|\mathbb{K}|$. This is the main difference between the bound in this theorem and in the previous result. This can be a major advantage when $|\zeta| \ll |\mathbb{K}|$ (e.g., in the monitoring problem). The disadvantage of this algorithm is that in practice tuning c might be difficult, since, typically, d is unknown.

6.3.4 An Upper Confidence Bound (UCB)-Based Policy for Switching Cost Consideration

To deal with switching costs, Le *et al.* proposed a policy in [LSZ14] that adapts the UCB2 policy in Sect. 2.4.2, denoted as *UCB2-Spanner*. The algorithm first plays each arm in the spanner ζ once. From there on, the decision time instances for arm selection are denoted by t_j, $j = 1, \ldots, J_n$, where $t_1 = |\zeta| + 1$ and J_n is the number of decision time instances up to time n. The quantities t_j, $j = 1, \ldots, J_n$ divide the time into epochs of length $l_j = t_{j+1} - t_j$, $j = 1, \ldots, J_n - 1$ to be defined next. At time t_j, the algorithm chooses

$$\mathbf{k}(t_j) = \arg\max_{\mathbf{k} \in \zeta} V_{\mathbf{k}}(t_j - 1), \tag{6.15}$$

where

$$V_{\mathbf{k}}(t_j - 1) = \hat{\mu}_{\mathbf{k}}(t_j - 1) + \sum_{i \in \mathcal{M}_{\mathbf{k}}} \sqrt{\frac{\rho \ln t_j}{N_i(t_j - 1)}}, \tag{6.16}$$

$$\hat{\mu}_{\mathbf{k}}(t_j - 1) = \hat{\theta}(t_j - 1)^\top \phi_{\mathbf{k}}. \tag{6.17}$$

Then, arm $\mathbf{k}(t_j)$ is played l_j times. From (6.16), we see that the choice of arm at time t_j is determined by two factors, namely, the estimated payoff in (6.17) (an approximation of the expected payoff in (6.6)) and a confidence bound determined by the number of times a component has been observed. Maximizing the first term gives exploitation, while the second term reflects the need for exploration. The choice in (6.15) trades off exploitation and exploration in order to reduce the sampling regret in (6.8).

Let $I(t_j) = \arg\min_{m \in \mathcal{M}_{\mathbf{k}(t)}} N_m(t_j - 1)$ (ties can be broken, say, in favor of the smallest index). Each component i is associated with an epoch counter $r_i(t)$ initialized to zero. At time t_j, the epoch counter of component $I(t_j)$ is updated as $r_{I(t_j)}(t_j) = r_{I(t_j)}(t_j - 1) + 1$, and remains the same for the rest of the epoch. The epoch length is given by $l_j = \tau(r_{I(t_j)}(t_j)) - \tau(r_{I(t_j)}(t_j) - 1)$. In other words, the epoch length is associated with the epoch counter of the least visited component. Note that the parameters are updated using (6.7) in each time slot after every observation, while the decisions are changed at the end of epochs only. Theorem 6.4 summarizes the regret bound of the UCB algorithm. Compared with the results in Theorem 6.2, the main difference lies in the term related to switching costs.

Theorem 6.4 (Regret bound of UCB2-Spanner) *Choose any ρ that satisfies $\rho > 1/1.99$. Then, there exists a constant $C > 0$ (which may depend on ρ) such that for all $n \geq 1$, the expected regret of our algorithm satisfies*

$$Q_n^{\text{UCB-Spanner}} \leq 4M \Delta_{\max} \left(\max_{k:\Delta_k>0} \frac{M_k}{\Delta_k} \right)^2 \rho \ln n + M \log_{1+\alpha} \ln n + C, \quad (6.18)$$

where $\Delta_{\max} = \max_k \Delta_k$.

6.3.5 Approximate Learning Algorithms

As discussed in Sect. 6.2.3.2, the main challenge in applying the UCB or ε-greedy algorithms is the computation complexity. In the sniffer-channel assignment problem, even with complete information, finding the optimal assignment is NP-hard. Therefore, there is a need to devise a computational efficient learning algorithm, which has sublinear regret in time with respect to an offline approximation solution.

Consider the simple greedy algorithm (GREEDY) that incrementally picks an unassigned sniffer and chooses the channel that maximizes the average number of remaining users it monitors. It has been shown in [Chh+10] to be $\frac{1}{2}$-approximate.

Denote $\mathbf{k}^g$ the assignment (or arm) chosen by GREEDY with complete information. We define the regret of policy π relative to GREEDY as,

$$R_g^\pi(n) = \mathbb{E}\left[\sum_{t=1}^n \left\{ \phi_{\mathbf{k}^g}^T \theta - \phi_{\mathbf{k}(t)}^T \theta \right\} \right]. \quad (6.19)$$

Note that by the property of GREEDY, $\frac{1}{2} \max_{\mathbf{k} \in \mathbb{K}} \phi_\mathbf{k}^T \theta - \phi_{\mathbf{k}^g}^T \theta \leq 0$. Thus, we have

$$R_g^\pi(n) \geq \mathbb{E}\left[\sum_{t=1}^n \left\{ \frac{1}{2} \max_{\mathbf{k} \in \mathbb{K}} \phi_\mathbf{k}^T \theta - \phi_{\mathbf{k}(t)}^T \theta \right\} \right].$$

We first rewrite GREEDY in terms of θ. GREEDY proceeds in m stages. Let the arm chosen by stage l be $\mathbf{k}_l = (\langle i_1, k_{i_1} \rangle, \langle i_2, k_{i_2} \rangle, \ldots, \langle i_l, k_{i_l} \rangle)$. Let $\oplus$ denote concatenation. In the $l + 1^{st}$ stage, GREEDY picks sniffer i_{l+1} and assign it channel $k_{i_{l+1}}$ if and only if the following condition holds

$$i_{l+1} = \arg \max_{j \in \mathscr{S}/\{i_1,i_2,\ldots,i_l\}} \max_{c=1,2,\ldots,K} (\phi_{\mathbf{k}_l \oplus \langle j,c \rangle}^T \theta - \phi_{\mathbf{k}_l}^T \theta), \quad (6.20)$$

Algorithm 6.1: The ε- GREEDY- APPROX algorithm

---

1 Define the sequence $\varepsilon_t \in (0, 1], t = 1, 2, \ldots$ by

$$\varepsilon_t \overset{def}{=} \min\left\{1, \frac{c}{t}\right\};$$ (6.22)

2 **foreach** $t = 1, 2, \ldots$ **do**
3   Let i_t be the arm picked by GREEDY;
4   With probability $1 - \varepsilon_t$ play i_t and with probability ε_t play a random spanner arm;
5   Observe the feedback and update the estimation of parameters using (6.7).

---

and

$$k_{i_{l+1}} = \arg \max_{c=1,2,\ldots,K} (\phi_{k_l \oplus \langle i_{l+1}, c\rangle}^T \theta - \phi_{k_l}^T \theta).$$ (6.21)

In the $l + 1^{st}$ stage, the total number of choices are $K(m - l)$. GREEDY needs to perform $K^2(m - l)(2^m - 1)$ multiplications to compute the expected payoffs and make $K(m - l)$ comparisons to find the optimal. Thus, the total computation complexity is $O(m^2 K^2(2^m - 1))$ compared to $O(K^m)$ time to determine the optimal assignment through enumeration. Note, the actual computation time is generally less by only considering the nonzero entries in the arm feature.

6.3.5.1 The ε- GREEDY APPROX **Algorithm**

The ε-greedy-approx algorithm uses GREEDY as a subroutine. The algorithm is summarized in Algorithm 6.1.

To establish the regret bound of the ε- GREEDY- APPROX algorithm, we first introduce the following lemma.

Lemma 6.1 *Given θ, there exists a nonempty $B \in [0, 1]^{K \cdot (2^m - 1)}$ centered at θ such that $\forall \hat{\theta} \in B$, the actions chosen by Greedy $\mathbf{k}^g(\hat{\theta})$ is identical to $\mathbf{k}^g(\theta)$.*

Lemma 6.1 implies that as long as $\hat{\theta}$, the estimate of θ, is sufficiently close to θ, the channel assignment of GREEDY is identical. The regret bound of the ε- GREEDY- APPROX algorithm is summarized in Theorem 6.5.

Theorem 6.5 (Regret bound of ε- GREEDY- APPROX) *Let*

$$\varepsilon_n = \min\left\{1, \frac{c}{n}\right\}, n > |\zeta|,$$ (6.23)

where $c > 0$ is a tuning parameter. Then, assuming that $c > \min(10|\zeta|, \frac{4|\zeta|}{d^2})$, where $d = \min_{k:\Delta_k > 0} \Delta_k$, the expected regret of ε- GREEDY- APPROX satisfies

$$R_g^{\varepsilon\text{-greedy-approx}}(n) \leq c \ln(n + 1) + O(1).$$ (6.24)

6.3.5.2   The EXP3- APPROX **Algorithm**

To further improve computation efficiency, a *multi-agent* multi-arm learning strategy
called EXP3- APPROX is proposed in [ZLH13]. This algorithm can also be implemented distributively. The basic idea is to have m agents, each corresponding to one
stage of GREEDY. Each agent maintains its own set of parameter estimates and makes
decision accordingly by treating decisions from agents prior to itself as a blackbox.
Learning of the agents is coupled by the way that the payoffs are partitioned among
the agents. Since the payoff of the kth agent depends on the actions of the previous
$k - 1$ agents, the rewards over time of an individual agent (with the exception of the
first agent) are no longer identically distributed. However, by viewing the reward as
assigned by an adversary who acts in accordance to the first $k - 1$ agents, we can
leverage policies for non-stochastic MAB on the kth agent.

   Consider m agents $\mathcal{E}_1, \mathcal{E}_2, \ldots, \mathcal{E}_m$. Agent $\mathcal{E}_i$ keeps a weight matrix W^i of
dimension $M \times K$. At time t, agent $\mathcal{E}_1, \mathcal{E}_2, \ldots, \mathcal{E}_m$ chose their respective action
in the action space $\mathscr{S} \times \mathscr{K}$ according to a modified version of the EXP3.1
algorithm [Aue+02][4]. It is assumed that agent $\mathcal{E}_l$ knows about the decision of agent
$\mathcal{E}_1$ through $\mathcal{E}_{l-1}$, namely, $\mathbf{k}_l = (\langle i_1, k_{i_1} \rangle, \langle i_2, k_{i_2} \rangle, \ldots, \langle i_{l-1}, k_{i_{l-1}} \rangle)$.

   Algorithm 6.2 summarizes the inner EXP3.1 algorithm executed by agent $\mathcal{E}_l$ at
time t. In line 5, the sniffer s and its associated channel k are chosen by probability
$p_{s,k}$. $p_{s,k}$ is a weighted average of two parts. The first part picks the tuple $\langle s, k \rangle$
proportionally to its weight in matrix W^l; the second part picks a tuple randomly.
The probability of picking a randomly action is γ, which decreases with the round
r_l over time.

   It has been proven in [Aue+02], EXP3.1 can achieve a uniform regret bound of
$O(\sqrt{nK' \ln K'})$ at time n with respect to an optimal policy using a single action,
where K' is the number of arms. Compared to EXP3.1 in [Aue+02], the main differences in the modified EXP3.1 is the consideration of only a subset of actions based
on the previous agents' decisions. This requires normalization of $w^l_{s,k}$ accordingly.
Similarly, we can adapt the EXP3.P algorithm in [Aue+02], which has less variances
in the regrets.

   The EXP3- APPROX (Algorithm 6.3) uses the modified EXP3 as a subroutine for
each agent. It is easy to see that EXP3- APPROX has a computation complexity of
$O(mKM)$ per update.

   To see the regret bound of EXP- APPROX, we need to first establish a few lem-
mas. Let $\tilde{g}$ be GREEDY when decisions are made with some additive errors. More
specifically, $\tilde{g}$ chooses $a_{j+1} = \langle i_{j+1}, k_{i_{j+1}} \rangle$ in stage $j + 1$ if and only if

$$f(\mathbf{k}_j^{\tilde{g}} \oplus a_{j+1}) - f(\mathbf{k}_j^{\tilde{g}}) \geq \max_{\langle s,c \rangle} \{ f(\mathbf{k}_j^{\tilde{g}} \oplus \langle s, c \rangle) - f(\mathbf{k}_j^{\tilde{g}}) \} - \varepsilon_{j+1}, \qquad (6.25)$$

where ε_{j+1} is the additive error in stage $j + 1$.

---

[4]EXP3.1 is a variation of Exp3 algorithm introduced in Chap. 4.

Algorithm 6.2: The MODIFIED EXP3.1 algorithm

    input   : $\langle i_j, k_{i_j} \rangle, j = 1, 2, \ldots, l-1$

1 ;

    initialization: Let $t = 1$, and $\hat{G}_{i,j}(1) = 0$ for $i = 1, 2, \ldots, M, j = 1, 2, \ldots, K$;

        $w_{s,k}^l = 1, s \in \mathscr{S}, k \in \mathscr{K}; r_l = 0$;

2 $C = \mathscr{S} \setminus \{i_1, i_2, \ldots, i_{l-1}\}$;

3 $g = MK \ln MK / (e-1)4^{r_l}$;

4 $\gamma = \min\left\{1, \frac{1}{2l}\right\}$;

5 $p_{s,k} = (1 - \gamma)\frac{w_{s,k}^l}{\sum_{(i,j) \in C} w_{i,j}^l} + \frac{\gamma}{MK}, \forall s \in \mathscr{S}, k \in \mathscr{K}$;

6 **repeat**

7 | Draw (i_l, k_{i_l}) randomly according to the probabilities $p_{s,k}$;

8 **until** $i_l \notin C$;

9 Receive reward $x_{i_l, k_{i_l}}(t)$;

10 $\hat{x}_{i_1, k_{i_l}} = \frac{x_{i_l, k_{i_l}}(t)}{p_{i_l, k_{i_l}}} \sum_{s \notin C, k \in \mathscr{K}} p_{s,k}$;

11 $w_{i_1, k_{i_l}}^l = w_{i_1, k_{i_l}}^l \exp\left(\frac{\gamma \hat{x}_{i_l, k_{i_l}}}{MK}\right)$;

12 $\hat{G}_{i_l, k_{i_l}}(t+1) = \hat{G}_{i_l, k_{i_l}}(t) + x_{i_l, k_{i_l}}(t)$;

13 $t = t + 1$;

14 **if** $\max_{i,j} \hat{G}_{i,j}(t) > g - MK/\gamma$ **then**

15   $\lfloor$ $r_l \leftarrow r_l + 1$;

Algorithm 6.3: The EXP3- APPROX algorithm

1 **foreach** $n = 1, 2, \ldots, T$ **do**

2   **foreach** $l = 1, 2, \ldots, m$ **do**

3     $\gamma = \min\{1, \sqrt{\frac{(M-(u-1))K \ln(M-(u-1))K}{(e-1)T}}\}$;

4   Run modified EXP3.1 to select an action $\langle i_l, k_{i_l} \rangle$;

5   Play $\mathbf{k} = (\langle i_1, k_{i_1} \rangle, \langle i_2, k_{i_2} \rangle, \ldots, \langle i_m, k_{i_m} \rangle)$ and observe
      $Y_n^o = (U_{i_1, k_{i_1}}, U_{i_2, k_{i_2}}, \ldots, U_{i_m, k_{i_m}})$;

6   **foreach** $l = 1, 2, \ldots, m$ **do**

7   Feedback $x_{i_l, k_{i_l}} = \left| \bigcup_{j=1}^{l} U_{i_j, k_{i_j}} - \bigcup_{j=1}^{l-1} U_{i_j, k_{i_j}} \right|$ to agent l;

8   Agent $\mathscr{E}_l$ updates W^l;

Lemma 6.2 $\tilde{g}$ *satisfies*

$$f(\mathbf{k}^{\tilde{g}}) \geq \frac{1}{2}\left(\max_{\mathbf{k} \in \mathbb{K}} f(\mathbf{k}) - \sum_{j=1}^{p} \varepsilon_j\right).$$

Lemma 6.2 states that with additive errors, GREEDY can achieve a utility no less than half of that of the optimal algorithm minus the sum of the additive errors.

Let r_i be the expected regret experienced by agent $\mathscr{E}_i$ and let $R = \sum_{i=1}^{m} r_i$. The following lemma relates the regret experienced by the agents to the regret of the original online problem.

Lemma 6.3 $R_g^{EXP3-approx}(n) < R/2.$

Note that in EXP3- APPROX, the feedback $x_{i_l, k_{k_l}}$ to agent $\mathscr{E}_l$ at time t satisfies $\mathbb{E}\left[x_{i_l, k_{k_l}}\right] = f(\mathbf{k}_{l-1} \oplus \langle i_l, k_{k_l} \rangle) - f(\mathbf{k}_{l-1})$ that depends on the choices of agents $\mathscr{E}_1, \mathscr{E}_2, \ldots, \mathscr{E}_{l-1}$ and is independent of the choices of agent $\mathscr{E}_t$ in the previous rounds. Therefore, each agent faces an adversarial multi-arm bandit [Aue+02]. Using the same arguments as the proof of Theorem 12 in [SG08], we have:

Theorem 6.6 *For finite time horizon n, algorithm* EXP3- APPROX *has an expected regret,*

$$\mathbb{E}\left[R_g^{EXP3-approx}(T)\right] = O(\sqrt{nMK\ln MK}).$$

6.4 Numerical Results

In this section, we present some numerical simulation results to illustrate the performances of the different algorithms discussed in this chapter.

In the simulations, wireless users are placed randomly in a 2-D plane. The area is partitioned into hexagon cells with circumcircle of radius 86 m. Each cell is associated with a base station operating in a channel (and so are the users in the cell). The channel to base station assignment ensures that *no neighboring cells use the same channel*. Sniffers are deployed in a grid formation separated by distance 100 m, with

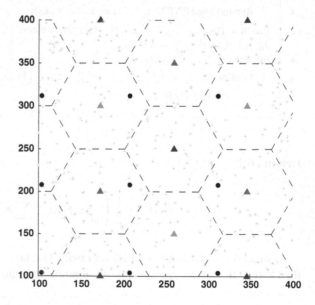

Fig. 6.1 Hexagonal layout with users ('+'), sniffers (solid dots), and channels of each cell (in different *colors*)

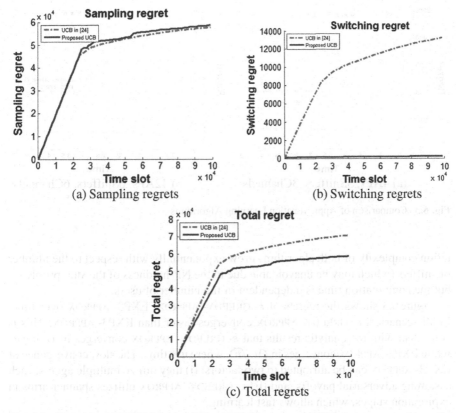

Fig. 6.2 Comparison of UCB-based policies, UCB-Spanner and UCB2-Spanner over the configuration of 12 APs using three channels, six sniffers, and 328 users

a coverage radius of 120 m. A snap shot of the synthetic deployment is shown in Fig. 6.1. The transmission probability of users is selected uniformly from [0, 0.06], resulting in an average busy probability of 0.2685 in each cell. We vary the total number of orthogonal channels from 3 to 6,[5] the number of cells from 4 to 12, and the number of sniffers from 3 to 6.

Figure 6.2 shows the sampling and switching regrets of the UCB-Spanner in Sect. 6.3.2 and the UCB2-Spanner algorithm in Sect. 6.3.4 over time. It can be seen that, the UCB2-Spanner that accounts for switching costs achieves similar sampling regrets as that of UCB-Spanner (Fig. 6.2a) even though the new method explores less often due to the use of epochs. When comparing the switching regrets (Fig. 6.2) and total regrets (Fig. 6.2c), UCB2-Spanner clearly outperforms UCB-2, as the later does not explicitly consider switching costs. The growth of the switching regrets in UCB2-Spanner is much slower—roughly at the rate of $O(\log \log n)$, while that of UCB-Spanner grows roughly as $O(\log n)$. Finally, it should be noted that the compu-

---

[5] In 802.11a networks, there are 8 orthogonal channels in 5.18–5.4 GHz, and one in 5.75 GHz.

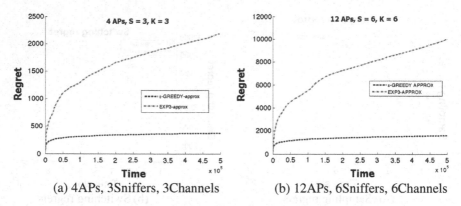

Fig. 6.3 Comparison of Approximation Learning Algorithms

tation complexity of both algorithms grow exponentially with respect to the number of sniffers (which may be unavoidable due to the NP-hardness of the MEC problem), but the computation time is independent of the number of users.

Figure 6.3 shows the regrets of ε- GREEDY- APPROX, EXP3- APPROX over time. In all scenarios, ε- GREEDY- APPROX converges faster than EXP3- APPROX. This is consistent with the analytic results that ε- GREEDY- APPROX converges in $O(\log n)$ while EXP3- APPROX converges in $O(\sqrt{n})$, where n is time. The slow convergence of EXP3- APPROX can be attributed to two factors: (i) they utilize multiple agents, each assuming adversarial payoffs, and (ii) ε- GREEDY- APPROX utilizes spanner arms in exploration stages, which allows fast learning.

6.5 Summary

In this chapter, we discussed the stochastic MAB formulation of the sniffer-channel assignment problem. It allows efficient passive monitoring of multichannel wireless networks. The problem bares some similarity to the spectrum sensing and access problem discussed in Chap. 5. In both problems, the utility of channel assignment depends on activities of users of the channels. However, since network monitoring is done passively, there is no "channel access" involved. The techniques developed in this chapter extend the stochastic MAB framework to address practical concerns such as computation complexity and switching costs in geographically distributed wireless networks. The EXP3- APPROX algorithm can be applied to nonstationary wireless channels as its inner block EXP3 was originally designed for adversarial settings.

References

[ABF02]   P. Auer, N. C. Bianchi, and P. Fischer. "Finite-time Analysis of the Multiarmed Bandit Problem". In: *Mach. Learn.* 47.2-3 (May 2002), pp. 235–256. ISSN: 0885-6125.

[AMT10]   A. Anandkumar, N. Michael, and A. K. Tang. "Opportunistic Spectrum Access with Multiple Users: Learning under Competition". In: *Proc. of IEEE INFOCOM.* San Diego, CA, 2010, pp. 1–9.

[ASZ11]   P. Arora, C. Szepesvári, and R. Zheng. "Sequential Learning for Optimal Monitoring of Multi-Channel Wireless Networks". In: *INFOCOM.* Shanghai, China, Apr. 2011, pp. 1152–1160.

[Aue+02]   P. Auer et al. "The Nonstochastic Multiarmed Bandit Problem". In: *SIAM J. Comput.* 32.1 (Feb. 2002), pp. 48–77.

[AVW87]   V. Anantharam, P. Varaiya, and J. Walrand. "Asymptotically efficient allocation rules for the multiarmed bandit problem with multiple plays-Part I: I.I.D. rewards". In: *Automatic Control, IEEE Transactions on* 32.11 (Nov. 1987), pp. 968–976. ISSN: 0018-9286.

[Che+06]   Y. C. Cheng et al. "Jigsaw: Solving the Puzzle of Enterprise 802.11 Analysis". In: *SIGCOMM.* Pisa, Italy, Sept. 2006, pp. 39–50.

[Chh+10]   A. Chhetri et al. "On Quality of Monitoring for Multi-channel Wireless Infrastructure Networks". In: *MobiHoc.* Chicago, IL, 2010, pp. 111–120.

[CK04]   C. Chekuri and A. Kumar. "Maximum Coverage Problem with Group Budget Constraints and Applications". In: *APPROX.* Cambridge, MA, Aug. 2004, pp. 72–83.

[CZ09]   Arun Chhetri and Rong Zheng. "WiserAnalyzer: A Passive Monitoring Framework for WLANs". In: *Proceedings of the 5th International Conference on Mobile Ad-hoc and Sensor Networks (MSN).* 2009.

[EGE02]   Jeremy Elson, Lewis Girod, and Deborah Estrin. "Fine-grained network time synchronization using reference broadcasts". In: *SIGOPS Oper. Syst. Rev.* 36.SI (2002), pp. 147–163. ISSN: 0163-5980.

[GNU]   GNURadio. "GNU radio project [ONLINE]". http://gnuradio.org/trac.

[LLC]   Ettus Research LLC. USRP2 - *The Next Generation of Software Radio Systems [ONLINE].* http://www.ettus.com/downloads/ettus_ds_usrp2_v2.pdf.

[LSZ14]   Thanh Le, Csaba Szepesvari, and Rong Zheng. "Sequential learning for multi-channel wireless network monitoring with channel switching costs". In: *IEEE Transactions on Signal Processing* 62.22 (2014), pp. 5919–5929.

[LZ10]   K. Liu and Q. Zhao. "Distributed Learning in Multi-Armed Bandit With Multiple Players". In: *IEEE Transactions on Signal Processing* 58.11 (Nov. 2010), pp. 5667–5681. ISSN: 1053-587X.

[Rod+05]   M. Rodrig et al. "Measurement-based Characterization of 802.11 in a Hotspot Setting". In: *Proceedings of the 2005 ACM SIGCOMM workshop on Experimental approaches to wireless network design and analysis.* Philadelphia, PA, USA, Aug. 2005, pp. 5–10. ISBN: 1-59593- 026-4.

[SB08]   Ashish Sharma and Elizabeth M. Belding. "FreeMAC: Framework for Multi-Channel MAC Development on 802.11 Hardware". In: *Proceedings of the ACM Workshop on Programmable Routers for Extensible Services of Tomorrow.* 2008.

[SB09]   D. H. Shin and S. Bagchi. "Optimal Monitoring in Multi-Channel Multi-Radio Wireless Mesh Networks". In: *MobiHoc.* New Orleans, LA, May 2009, pp. 229–238.

[SG08]   M. J. Streeter and D. Golovin. "An Online Algorithm for Maximizing Submodular Functions". In: *NIPS.* Vancouver, Canada, Dec. 2008, pp. 1577–1584.

[Yeo+05]   J. Yeo et al. "An Accurate Technique for Measuring the Wireless Side of Wireless Networks". In: *The 2005 workshop on Wireless traffic measurements and modeling.* Seattle, WA, June 2005, pp. 13–18. ISBN: 1-931971-33-1.

[YYA04]   J. Yeo, M. Youssef, and A. Agrawala. "A Framework for Wireless LAN Monitoring and its Applications". In: *WiSe '04: Proceedings of the 3rd ACM workshop on Wireless security.* Philadelphia, PA, Oct. 2004, pp. 70–79. ISBN: 1-58113-925-X.

[ZLH13]   Rong Zheng, Thanh Le, and Zhu Han. "Approximate online learning for passive mon-
          itoring of multi-channel wireless networks". In: *Proceedings of the IEEE INFOCOM
          2013, Turin, Italy, April 14-19, 2013. 2013*, pp. 3111–3119.

Chapter 7
Online Routing in Multi-hop Wireless Networks

In this chapter, we investigate the routing problem in multi-hop wireless networks, whereby link qualities are unknown and time-varying due to fast-fading wireless channels. Each flow has to select a path from its source to its destination from a set of candidate paths so as to minimize the expected end-to-end routing cost. The challenge for this problem is twofold. First, the routing cost for each candidate path is unknown and has to be learned sequentially due to dynamic link qualities. Secondly, the routing cost over a link also depends on the number of flows passing through it, which is unknown since the routing decisions of flows are made online. We formulate this problem as an online learning problem based on the adversarial MAB framework, which accommodates both the arbitrary variation of link qualities and the unpredictable competition among flows. An online routing algorithm is presented based on an exponentially weighted average strategy, which is shown to converge to a set of correlated equilibria with vanishing regrets through theoretical analysis and simulations.

7.1 Introduction

We consider the routing problem in a multi-hop wireless network. Specifically, consider there are multiple flows in the network. Each flow is associated with a unique pair of source and destination nodes. In the conventional setting, the parameters (throughput, delay, packet loss ratio, etc.) of each link in the network are assumed to be known a priori. The problem of finding the shortest (or the minimum cost) path for each individual flows (known as shortest path routing) has been studied extensively in literature [Cad+13, Gao+05].

Unfortunately, due to the innate dynamic nature of wireless channels (e.g., fast-fading, shadowing, co-channel interference, etc.), the quality of wireless links changes over time under an unknown distribution. In addition, the routing cost of

© Springer International Publishing AG 2016   91
R. Zheng and C. Hua, *Sequential Learning and Decision-Making in Wireless Resource Management*, Wireless Networks, DOI 10.1007/978-3-319-50502-2_7

a link is also affected by the competing flows passing through the link, which is unknown since the routing decisions of flows are made online. Therefore, the link parameters can only be estimated (or observed) after the packets of the flow have been routed through the network, which lead to the challenging trade-off between exploration and exploitation. That is, on the one hand, the candidate routing paths should be explored as much as possible so that the optimal one can be found eventually. On the other hand, the knowledge about the link states should be exploited so that the paths with smaller cost can be selected.

This routing problem is often referred to the *online (or adaptive) shortest path routing problem* in literature [AK04, Gyö+07], which falls into the classic combinatorial Multi-armed Bandit (MAB) framework. In [GKJ12], the authors propose a LLC (learning with linear cost) policy for the adaptive routing problem, which is based on the general multi-armed bandit problem with linear cost defined on any constraint set. Using the LLC policy, each path between the source and destination is mapped as an action. By adopting Dijkstra's algorithm or Bellman–Ford algorithm, LLC can solve the online shortest path problem with polynomial computation time and achieve a regret of $O(N^4 \log T)$, where N is the number of edges in the network, and T is the time horizon. In [LZ12], the authors assume that the quality of each link varies over time according to a stochastic process with unknown distributions. After a path is selected, only the aggregated cost of all links along this path is observed, and the quality of each individual link is not observable. This problem is modeled as a multi-armed bandit with dependent arms, and an efficient online learning algorithm is proposed to minimize the regret in terms of both time and network size. In [He+13], the authors study the decoupled routing policy, that is, the source sends probing packets on one path and data packets on another, so that these two procedures can be optimized separately. It is proven that the decoupled routing policy reduces the regret from $(\log T)$ under coupled probing/routing policy as shown in [GKJ12] to a value constant in T. In [Tal+15], the authors study the online shortest path routing problem based on the stochastic MAB framework. Three types of routing policies are analyzed: source routing with semi-bandit feedback, source routing with bandit feedback, and hop-by-hop routing. It is proven that semi-bandit feedback significantly improves the performance, while hop-by-hop decisions do not achieve much performance gain. In [AK08], the online routing problem is studied under an adversarial bandit model in which the cost functions are chosen by an adversary and are treated as arbitrary bounded deterministic quantities. Algorithms were proposed to achieve regrets sublinear with respect to the time horizon and polynomial with respect to the network size.

In this chapter, we consider the online routing problem in a dynamic multi-hop wireless network, whereby the link costs are unknown and have to be learned sequentially. Moreover, we assume that multiple flows are traversing through the network, so that the link costs are also affected by the competing flows passing through the same link. As a result, the problem involves the modeling of the selfishness of individual flows as well. We formulate this problem as an online learning problem based on the adversarial MAB model. An exponentially weighted average strategy is adopted to design an online routing algorithm, which is guaranteed to converge to a set of correlated equilibria with vanishing internal regret. Simulation results demonstrate the superior performance of the proposed scheme.

The rest of this chapter is organized as follows. In Sect. 7.2, we introduce the system model and state the online routing problem. An online rouging algorithm is proposed in Sect. 7.3. Some numerical results are provided in Sect. 7.4 to illustrate the property of the proposed algorithm and finally conclusions are drawn in Sect. 7.5.

7.2 System Model and Problem Statement

7.2.1 Network Model

We consider the routing problem in a multi-hop wireless network as shown in Fig. 7.1, where the network topology can be modeled as a graph $G(V, E)$, V is the set of nodes, E is the set of links and $M = |E|$ is the total number of links. There are K flows in the network, and each flow k is associated with a pair of source node s_k and destination node d_k. Let Q_k denote the set of candidate paths from s_k to d_k, and $N_k = |Q_k|$ is the number of candidate paths for flow k.

The quality of a link $l \in E$ at time t is characterized by the link capacity $C_{l,t}$, which is a function of its perceived Sinal-to-Noise ratio (SNR), i.e., $C_{l,t} = f(SNR_l)$. For example, for AWGN channels, the link capacity can be upper bounded by Shannon's equation, that is, $C_{l,t} \leq B \log \left(1 + \frac{T_l G_l(t)}{N_0}\right)$, where B is the channel bandwidth in hertz, T_l is the transmit power, $G_l(t)$ is the instantaneous channel gain at time t, and N_0 is the noise power. Due to fast-fading and shadowing effects, the channel gain changes over time following an unknown distribution, so the link capacity is also time-varying. Here, we assume that $C_{l,1}, \ldots, C_{l,t}$ are i.i.d. random variables.

We focus on the delay experienced at a link, which depends on its capacity as well as the aggregate traffic passing through the link. In general, link delay can be derived based on the classic queueing models such as M/M/1, M/G/1, G/G/1, or the CSMA/CA model for the IEEE 802.11 protocol. For example, using the simple M/M/1 queueing model, the delay incurred at a link l can be given by:

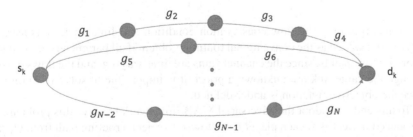

Fig. 7.1 Online routing problem

$$g_l(t) = \frac{1}{C_{l,t} - \lambda_l} = \frac{1}{B \log \left(1 + \frac{T_l G_l(t)}{N_0}\right) - \sum_{i \in F_l} \lambda_i}, \tag{7.1}$$

where F_l is the set of flows passing through link l, and $\lambda_l = \sum_{j \in F_l} \lambda_j$ is the aggregate traffic load at link l. The analysis can be adapted to consider other queueing models as well. In practice, the link delay should be measured online based on the data packets (or probe packets) transmitted over the link (or path), and then fed back to the source node to make routing decision.

7.2.2 Online Routing Problem

We assume that time is divided into slots with equal duration. Link costs are unknown at the beginning of each time slot, and thus the source node of each flow (viewed as a player) can only make its decision according to historical information. Specifically, at the beginning of each time slot t, a player k chooses a routing path $I_t^{(k)}$ from Q_k and sends its packets over the path. At the end of time slot t, the player can observe the cost of his selected routing path according to the feedback from the destination node.

The cost for a player k to select routing path $I_t^{(k)}$ at time t is defined as the sum of the delays over all links along path $I_t^{(k)}$, that is,

$$g^{(k)}(t, \mathbf{I}_t) = \sum_{l \in I_t^{(k)}} \frac{1}{B \log \left(1 + \frac{T_l G_l(t)}{N_0}\right) - \sum_{i \in F_l} \lambda_i}, \tag{7.2}$$

where $\mathbf{I}_t = (I_t^{(1)}, I_t^{(2)}, \ldots, I_t^{(K)})$ denotes the set of actions taken by all players at time t, i.e., $I_t^{(k)}$ corresponds to the selected path by player k.

The objective of player k is to find the optimal routing path to minimize its cumulative routing cost up to time n, that is

$$\underset{I_t^{(k)}}{\text{minimize}} \sum_{t=1}^{n} g^{(k)}(t, \mathbf{I}_t) \tag{7.3}$$

By incorporating the flow conservation condition and link capacity constraint, (7.3) can be solved as the conventional routing problem if all parameters are known a priori. Unfortunately, since the channel gains are time-varying, and the flows passing through the same link are unknown a priori, it is impossible to solve (7.3) directly since the objective function is under-defined.

To this end, we adopt the adversarial MAB framework to solve this problem. As introduced in Sect. 4.1, each player k is allowed to select a routing path from Q_k randomly according to the **routing probability vector** $P_{k,t} = \{p_{1,t}^{(k)}, \ldots, p_{i,t}^{(k)}, \ldots, p_{N_k,t}^{(k)}\}$, where $p_{i,t}^{(k)}$ denotes the probability that player k chooses routing path i at time t. The performance of this routing strategy can be measured by the external regret. Therefore, instead of solving (7.3), each player k tries to achieve per-round vanishing (zero average) regret as follows:

$$\lim_{n \to \infty} \frac{1}{n} (\sum_{t=1}^{n} \bar{g}^{(k)}(t, \mathbf{P}_t) - \min_{I_t^{(k)}=1,\ldots,N_k} \sum_{t=1}^{n} g^{(k)}(t, I_T^{(k)})) = 0, \qquad (7.4)$$

where n denotes the time horizon, $\mathbf{P}_t$ denotes the joint action profile of all players, and $\bar{g}^{(k)}(t, \mathbf{P}_t) = \sum_{i \in \mathcal{Q}_k} p_{i,t}^{(k)} g(t, i)$. This strategy translates to vanishing external regret when players use mixed strategies for selecting routing paths.

7.3 Algorithm

In this section, we present an algorithm for the online routing problem based on the bandit version of the internal regret minimizing algorithm with the exponentially weighted average strategy as introduced in Sect. 4.2.4. The basic idea of the algorithm is to select each path with a probability according to the accumulated internal regret caused by choosing this path. This algorithm has two parameters, γ_t and η_t. Setting $\gamma_t = t^{-\frac{1}{3}}$ and $\eta_t = \frac{\gamma_t^3}{N_k^2}$ guarantees the Hannan-consistency [CL06], which in turn leads to vanishing internal regrets.

The pseudocode of the algorithm for player k is shown in Algorithm 7.1. Initially, since no historical information is available, the player selects each action with equal probability and observes the routing cost (line 1). Then the algorithm proceeds in rounds. In each round, a probability $\mathbf{P}_{t-1}^{(k),i \to j}$ is constructed for each pair of actions i and j based on the probability distribution yielded in the previous round (line), to generate the *internal regret* $\tilde{r}_{i \to j,t-1}^{(k)}$ and the weighted parameter $\Delta_{i \to j,t}^{(k)}$ between actions i and j (line 6–7). Note that since each player can only observe the cost of the selected routing path but not those of others, the cost can only be estimated using unbiased estimator (7.7) (line 9), that is $E_t(\bar{g}(i, t)) = g(i, t)$, where E denotes the expectation operator. The action probability vector is updated by solving the fixed point equation in (7.8) (line 10), and then is used to obtain the final probability vector by the exponentially weighted average strategy (line 11). Finally, a routing path is selected based on the newly updated probability vector and the corresponding routing cost is observed (line 12).

The following result characterizes the performance of the proposed EWA-R algorithm.

Theorem 7.1 *EWA-R algorithm has a vanishing internal regret.*

Theorem 7.2 *If all players make routing decisions according to the EWA-R algorithm, the joint distribution of plays converges to the set of correlated equilibria.*

The proof of Theorem 7.1 can be found in Appendix. Theorem 7.2 follows naturally from Theorems 4.10 and 7.1.

Algorithm 7.1: EWA-R: Exponentially weighted average based routing algorithm

---

1 **Init:** *Set* $\gamma_t = t^{-\frac{1}{3}}$, $\eta_t = \frac{\gamma_t^3}{N_k^2}$, *and* $\mathbf{P}_1^{(k)} = (\frac{1}{N_k}, ..., \frac{1}{N_k})$;

2 Select a routing path $I_1^{(k)}$ according to $\mathbf{P}_1^k$ and observe the routing cost.

3 **for** $t = 2, ...,$ **do**

4 $\quad$ Let $\mathbf{P}_{t-1}^{(k)}$ be the probability distribution (mixed strategy) over N_K actions at time $t-1$,
$\quad$ i.e. $\mathbf{P}_{t-1}^{(k)} = (p_{1,t-1}^{(k)}, ..., p_{i,t-1}^{(k)}, ..., p_{j,t-1}^{(k)}, ..., p_{N_k,t-1}^{(k)})$.

5 $\quad$ Construct $\mathbf{P}_{t-1}^{(k),i\to j}$ as follows: set $p_{i,t-1}^{(k)}$ in $\mathbf{P}_{t-1}^{(k)}$ to zero, and replace $p_{j,t-1}^{(k)}$ with
$\quad$ $p_{i,t-1}^{(k)} + p_{j,t-1}^{(k)}$, other elements remain unchanged. We obtain
$\quad$ $\mathbf{P}_{t-1}^{(k),i\to j} = (p_{i,t-1}^{(k)}, ..., 0, ..., p_{i,t-1}^{(k)} + p_{j,t-1}^{(k)}, ..., p_{N_k,t-1}^{(k)})$.

6 $\quad$ Update

$$\Delta_{i\to j,t}^{(k)} = \frac{\exp(\eta_t \widetilde{r}_{i\to j,t-1}^{(k)})}{\sum_{(m\to l):m\neq l} \exp(\eta_t \widetilde{r}_{m\to l,t-1}^{(k)})} \tag{7.5}$$

7 $\quad$ where

$$\widetilde{r}_{i\to j,t-1}^{(k)} = \sum_{s=1}^{t-1} \widetilde{r}_{i\to j,s}^{(k)}$$
$$= \sum_{s=1}^{t-1} p_{i,s}^{(k)} (\widetilde{g}^{(k)}(s,j) - \widetilde{g}^{(k)}(s,i)), \tag{7.6}$$

8 $\quad$ **for** $i = 1, ..., N_k$ **do**

9

$$\widetilde{g}^{(k)}(s,i) = \begin{cases} \frac{g^{(k)}(s,\mathbf{I}_s)}{p_{i,s}^{(k)}} & \text{if } i = I_s^{(k)}. \\ 0 & \text{otherwise.} \end{cases} \tag{7.7}$$

10 $\quad$ Given $\Delta_{i\to j,t}^{(k)}$, solve the following fixed point equation to obtain $\mathbf{P}_t^{(k)}$:

$$\mathbf{P}_t^{(k)} = \sum_{(i,j):i\neq j} \mathbf{P}_t^{(k),i\to j} \Delta_{i\to j,t}^{(k)}. \tag{7.8}$$

11 $\quad$ Update the final probability distribution as:

$$\mathbf{P}_t^{(k)} = (1 - \gamma_t)\mathbf{P}_t^{(k)} + \frac{\gamma_t}{N_k} \tag{7.9}$$

12 $\quad$ Select a routing path $I_t^{(k)}$ from Q_k according to $\mathbf{P}_t^{(k)}$ and observe the routing cost.

7.4 Numerical Results

In this section, we evaluate the performance of the proposed online routing algorithm based on a simple network topology as depicted in Fig. 7.1. For simplicity, we set the same source and destination nodes for all flows (indicated as "s" and "d" in the figure), and three candidate paths are available for each flow. In the simulation, the

Fig. 7.2 Convergence of the routing probability vector

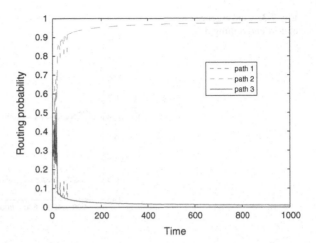

number of flows varies from 5 to 50. If the same link is shared by multiple flows, the link cost is determined according to (7.1). For a comparison. we also implement the **random** routing strategy that selects one of the paths randomly, and the **optimal** routing strategy that selects the best path assuming the channel information is known a priori.

First, we study the convergence of the EWA-R algorithm. In Fig. 7.2, we show the evolution of the routing probability vectors for a flow with time. It can be seen that as time t increases, the proposed EWA-R algorithm approaches to the correlated equilibrium. It should be noted that the exponentially weighted average strategy tends to keep the routing probability far away from 0, a sufficient time is needed for exploring each routing path. Therefore, the convergence rate is slowed down with the increase of time horizon.

Second, Fig. 7.3 compares the average delay achieved by the EWA-R and other two algorithms. As can be seen from the figure, the achieved delay by the optimal algorithm is almost constant since the link state information is assumed to be known and used for making routing decision. On the other hand, the delay achieved by the EWA-R algorithm approaches the optimal strategy quickly as time increases, which confirms the property of varnishing regret of the EWA-R algorithm.

Finally, we study the effect of the number of flows on the delay achieved by the EWA-R algorithm after it converges in Fig. 7.4. Since the delays achieved by the EAW-R algorithm and the optimal strategy are very close after convergence (see Fig. 7.3), we only compare the performance of the EWA-R scheme and the random schemes under different number of flows. It can be seen that the delay increases gradually with the increase of the number of flows in the network, and the EWA-R algorithm outperforms the random scheme under all conditions.

Fig. 7.3 Convergence of
end-to-end routing delay

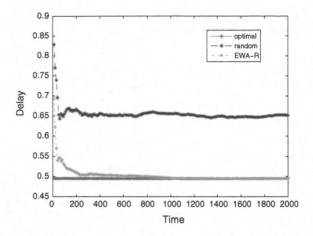

Fig. 7.4 End-to-end delay
versus number of flows

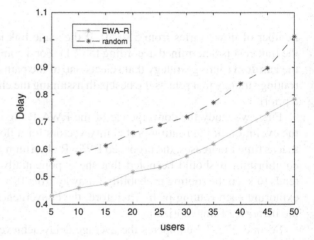

7.5 Summary

In this chapter, we studied the online routing problem in a dynamic wireless network. We applied the adversarial MAB framework for this problem, which captured not only the variations of link quality, but also the competition of flows on the same link. An online routing algorithm was designed based on the adversarial MAB framework, and its convergence property was provided. Simulation results demonstrated the convergence of the proposed EWA-R algorithm, and the comparison results with the random and optimal routing schemes were also provided to show the performance of the proposed algorithm.

Appendix

The proof of Theorem 7.1 in Sect. 7.3 is based on some auxiliary results and Theorem 6.9 of [CL06], reproduced below as Theorem 7.3.

Theorem 7.3 *Let* $\Phi(U) = \psi(\sum_{i=1}^{N} \phi(u_i))$, *where* $U = (u_1, \ldots, u_N)$. *Consider a selection strategy that selects action* I_t *at time* t *according to distribution* $P_t = \{p_{1,t}, p_{2,t}, \ldots, p_{N,t}\}$, *whose elements* $p_{i,t}$ *are defined as:*

$$p_{i,t} = (1 - \gamma_t) \frac{\phi'(R_{i,t-1})}{\sum_{k=1}^{N} \phi'(R_{i,t-1})} + \frac{\gamma_t}{N}, \tag{7.10}$$

where $R_{i,t-1} = \sum_{s=1}^{t=1} (g(s, i) - g(s, I_s))$.
If the following assumptions hold:

1. $\sum_{t=1}^{n} \frac{1}{\gamma_t^2} = o(\frac{n^2}{\ln n})$

2. *For all vector* $V_t = (v_{1,t}, \ldots, v_{n,t})$ *with* $\mid v_{i,t} \mid \leq \frac{N}{\gamma_t}$, *we have*

$$\lim_{n \to \infty} \frac{1}{\psi(\phi(n))} \sum_{t=1}^{n} C(V_t) = 0, \tag{7.11}$$

   where

$$C(V_t) = \sup_{U \in R^N} \psi' \left(\sum_{i=1}^{N} \phi(u_i) \right) \sum_{i=1}^{N} \phi''(u_i) v_{i,t}^2.$$

3. *For all vectors* $U_t = (u_{1,t}, \ldots, u_{n,t})$, *with* $u_{i,t} \leq t$,

$$\lim_{n \to \infty} \frac{1}{\psi(\phi(n))} \sum_{t=1}^{n} \gamma_t \sum_{i=1}^{N} \nabla_i \Phi(U_t) = 0, \tag{7.12}$$

4. *For all vectors* $U_t = (u_{1,t}, \ldots, u_{n,t})$, *with* $u_{i,t} \leq t$,

$$\lim_{n \to \infty} \frac{\ln n}{\psi(\phi(n))} \sqrt{\sum_{t=1}^{n} \frac{1}{\gamma_t^2} \left(\sum_{i=1}^{N} \nabla_i \Phi(U_t) \right)^2} = 0, \tag{7.13}$$

Then the selection strategy satisfies:

$$\lim_{n \to \infty} \frac{1}{n} \left(\max_{i=1,\ldots,N} \sum_{t=1}^{n} g(t, i) - \sum_{t=1}^{n} g(t, I_t) \right) = 0. \tag{7.14}$$

Proof of Theorem 7.1

In order to prove Theorem 7.1, we show that the selected parameters $\gamma_t = t^{-\frac{1}{3}}$ and $\eta_t = \frac{\gamma_t^3}{N^2}$ satisfy (1)–(4) in Theorem 7.3.

1. For $\gamma_t = t^{-\frac{1}{3}}$, we have

$$\sum_{t=1}^{n} \frac{1}{\gamma_t^2} = \sum_{t=1}^{n} t^{\frac{2}{3}} = H_n\left[\frac{-2}{3}\right]. \tag{7.15}$$

Then,

$$\lim_{n\to\infty} \frac{\ln n}{n^2} \sum_{t=1}^{n} \gamma_t^2 = \lim_{n\to\infty} \frac{\ln n}{n^2} H_n\left[\frac{-2}{3}\right] = 0. \tag{7.16}$$

2. For $\psi(x) = \frac{1}{\eta_t} \ln x$ and $\phi(x) = \exp(\eta_t x)$, we obtain

$$C(V_t) = \sup\left(\eta_t \sum_{i=1}^{} N v_{i,t}^2\right) = \frac{\eta_t N^3}{\gamma_t^2}. \tag{7.17}$$

Hence,

$$\lim_{n\to\infty} \frac{1}{\psi(\phi(n))} \sum_{t=1}^{n} C(V_t) = \lim_{n\to\infty} \frac{1}{n} \sum_{t=1}^{n} t^{-\frac{1}{3}}$$
$$= \lim_{n\to\infty} \frac{1}{n} H_n\left[\frac{1}{3}\right] = 0. \tag{7.18}$$

3. For $\Phi(U) = \frac{1}{\eta_t} \ln\left(\sum_{i=1}^{N} \exp(\eta_t u_i)\right)$, $\nabla_i \phi(U_t)$ is given by:

$$\nabla_i \phi(U_t) = \frac{\exp(\eta_t u_i)}{\sum_{i=1}^{N} \exp(\eta_t u_i)}. \tag{7.19}$$

Therefore,

$$\lim_{n\to\infty} \frac{1}{\psi(\phi(n))} \sum_{t=1}^{n} \gamma_t \sum_{i=1}^{N} \nabla_i \phi(U_t) =$$
$$\lim_{n\to\infty} \frac{1}{n} \sum_{t=1}^{n} t^{-\frac{1}{3}} \sum_{i=1}^{N} N \frac{\exp(\eta_t u_i)}{\sum_{i=1}^{N} \exp(\eta_t u_i)} = \tag{7.20}$$
$$\lim_{n\to\infty} \frac{1}{n} H_n\left[\frac{1}{3}\right] = 0.$$

4. By substituting (7.19) into (7.13), we obtain (4).

Therefore, (7.14) ensures since all assumptions (1)–(4) are held, which completes the proof of Theorem 7.1.

References

[AK04]   Baruch Awerbuch and Robert D. Kleinberg. "Adaptive Routing with End-to-end Feed-back: Distributed Learning and Geometric Approaches". In: *Proceedings of the Thirty-sixth Annual ACM Symposium on Theory of Computing*. STOC '04. 2004, pp. 45–53.

[AK08]   Baruch Awerbuch and Robert Kleinberg. "Online Linear Optimization and Adaptive Routing". In: *J. Comput. Syst. Sci.* 74.1 (Feb. 2008), pp. 97–114. ISSN: 0022-0000.

[Cad+13]   F. Cadger et al. "A Survey of Geographical Routing in Wireless Ad- Hoc Networks". In: *Communications Surveys Tutorials, IEEE* 15.2 (Second 2013), pp. 621–653. ISSN: 1553-877X.

[CL06]   N. Cesa-Bianchi and G. Lugosi. "Prediction, Learning, and Games". In: *Cambridge University Press*. 2006.

[Gao+05]   Jie Gao et al. "Geometric spanners for routing in mobile networks". In: *Selected Areas in Communications, IEEE Journal on* 23.1 (Jan. 2005), pp. 174–185. ISSN: 0733-8716.

[GKJ12]   Yi Gai, Bhaskar Krishnamachari, and Rahul Jain. "Combinatorial Network Optimization with Unknown Variables: Multi-armed Bandits with Linear Rewards and Individual Observations". In: *EEE/ACM Trans. Netw.* 20.5 (Oct. 2012), pp. 1466–1478. ISSN: 1063-6692.

[Gyö+07]   Andráas György et al. "The On-Line Shortest Path Problem Under Partial Monitoring". In: *J. Mach. Learn. Res.* 8 (Dec. 2007), pp. 2369– 2403. ISSN: 1532-4435.

[He+13]   Ting He et al. "Endhost-based shortest path routing in dynamic networks: An online learning approach". In: *IEEE INFOCOM'13*. Apr. 2013, pp. 2202–2210.

[LZ12]   Keqin Liu and Qing Zhao. "Adaptive shortest-path routing under unknown and stochastically varying link states". In: *Modeling and Optimization in Mobile, Ad Hoc and Wireless Networks (WiOpt), 2012 10th International Symposium on*. May 2012, pp. 232–237.

[Tal+15]   M. S. Talebi et al. "Stochastic Online Shortest Path Routing: The Value of Feedback". 2015.

Chapter 8
Channel Selection and User Association in WiFi Networks

In this chapter, we consider the emerging deployment of WiFi networks in sports and entertainment venues characterized by high-density, large capacity, and real-time service delivery. Due to extremely high user density, channel allocation and user association should be carefully managed so that co-channel inference can be mitigated. Channel selection and user association (CSUA) is formulated as an adversarial MAB problem, which captures not only the uncertainty of channel states, but also the selfishness of individual stations (STAs) and access points (APs). An exponentially weighted average strategy is adopted to design an online algorithm for this problem, which is guaranteed to converge to a set of correlated equilibria with vanishing regrets. Simulation results show the convergence of the proposed algorithm and its performance under different settings.

8.1 Introduction

To meet the growing demands for data service due to the rapid penetration of mobile devices (smart phones, iPad, laptop, etc.), WiFi networks have been selected by many service provider as a cost-effective 3G/4G offloading solution in public places such as shopping malls, bars, hotels, and public squares. In sports and entertainment venues, the availability of WiFi networks is also seen as an opportunity to improve the experience of customers and pursue revenue growth opportunities. Unlike common office networks, most stadiums have tens of thousands of seats, user density will be hundreds of times greater than in common office networks. For example, for a 50,000-seat stadium, around 400 APs would be needed to provide sufficient capacity and coverage throughout the venue area [C11]. Therefore, WiFi networks in these venues are characterized by high-density, large capacity, and real-time service delivery, making its deployment and management complicated. Recognizing the challenges arising in high-density WiFi networks, the IEEE 802.11 working group has formed a new study group with the objective to define a highly efficient WLAN

© Springer International Publishing AG 2016
R. Zheng and C. Hua, *Sequential Learning and Decision-Making in Wireless Resource Management*, Wireless Networks, DOI 10.1007/978-3-319-50502-2_8

standard, which was approved as 802.11ax project by the IEEE standard board in March 2014 [IEE14].

The main limiting factor in designing high-density networks is co-channel interference since the number of available orthogonal channels is far less than the number of neighboring APs in this case. Therefore, channels should be carefully assigned to APs so that co-channel interference from neighboring APs can be mitigated. The so-called channel assignment problem has been studied extensively in literature. For example, in [Yu+07], the authors formulate AP channel allocation as a min–max optimization problem with the objective of minimizing the channel utilization of the AP with the maximum traffic load and propose a dynamic radio channel allocation strategy. Based on the same framework, a channel allocation strategy is proposed for WLANs with multiple data rates [YMS08]. The channel assignment algorithm proposed in [Hai+08] aims to maximize the signal-to-interference ratio at user level. In [Yua+13], the channel allocation problem is reduced to a width allocation problem, which is then formulated as a noncooperative game whereby the desired allocation result corresponds to the Nash equilibrium. A comprehensive survey on the channel allocation strategy for WLANs can be found in [CHD10].

On the other hand, due to high user density, the number of stations (STAs) associated with an AP can be dozens or even hundreds and may distribute unevenly geographically. The performance of STAs served by overly loaded APs degrades dramatically due to the CSMA/CA based channel access mechanism. To address this problem, many user association schemes have been proposed in literature that attempt to balance traffic load across APs. In [Erc08], the authors analyze the convergence and stability of a user association strategy based on the congestion game theoretical model. The authors in [BHL04] propose a solution to user association so that the max–min fair bandwidth allocation is ensured. Recently, user association and channel allocation have been jointly considered to achieve better performance. For example, in [Kau+07], an algorithm is designed for this joint optimization problem based on the Gibbs sampler technique. In [XHH11], the channel allocation and user association problem is modeled as a noncooperative game, and the proposed algorithm is proven to converge to the Nash equilibrium.

Due to the dynamics of wireless channels (e.g., fast-fading, shadowing, and co-channel interference, etc.), the channel state information is unknown a priori. The problem of joint channel selection and user association is nontrivial since the co-channel interference and contention level can only be estimated (or observed) after the decisions have been made, which leads to the challenging trade-off between exploration (learning the statistics of individual channels and APs) and exploitation (utilizing the best channel and AP). Moreover, interference and contention also depend on the decision of other APs and STAs, which are unknown since their decisions are made online. In this chapter, we adopt the adversarial MAB framework to address these challenging issues, which handles not only the uncertainty of channel states, but also the selfishness of individual STAs and APs. Online algorithms are designed based on the exponentially weighted average strategy, which is guaranteed to converge to a set of correlated equilibria with vanishing regret. The main novelty in the proposed solution lies in its decentralized nature where individual APs and STAs in multiple contention domains make decisions entirely based on local measurements.

The rest of chapter is organized as follows. In Sect. 8.2, we introduce the system model and formally define the CSUA problem. In Sect. 8.3, the adversarial MAB framework is introduced for the CSUA problem and online algorithms are designed accordingly. Simulation results are provided in Sect. 8.4 to evaluate the performance of the proposed scheme and finally we conclude this chapter in Sect. 8.5.

8.2 System Models and Problem Statement

8.2.1 Network Model

We consider a wireless network consisting of a set of APs $\mathscr{A} = \{1, \cdots, A\}$ and a set of stations (STAs) $\mathscr{S} = \{1, \cdots, S\}$ based on the IEEE 802.11 protocol (Fig. 8.1). Let A_s denote the set of APs who are within the communication range of STA s. Similarly, S_a denotes the set of STAs associated with AP a.

Each AP can operate in a set of M orthogonal channels, and all STAs associated with an AP should operate in the same channel as the AP. Let $\mathscr{N}_a$ denote the set of neighboring APs who are in the contention range of AP a, which will lead to co-channel interference to AP a if they are operating in the same channel.

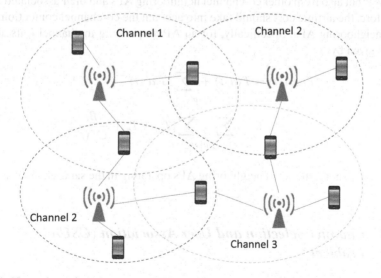

Fig. 8.1 Network topology for a dense WiFi network

8.2.2 Airtime Cost

We adopt the *airtime cost* as a measure of the delay between a pair of AP and STA, which was first proposed as a routing metric for 802.11s wireless mesh networks [IEE08]. Specifically, for a STA s and its associated AP a, the airtime cost is defined as follows:

$$T_s^a(t) = T_{ca} + T_p + \frac{B_t}{C_s^a(t)}, \tag{8.1}$$

where T_{ca} is the channel access overhead, T_p is the protocol overhead, B_t is the number of bits in the data frame at t, and $C_s^a(t)$ is the transmission rate between AP a and STA s at time t. Under saturated condition, T_{ca}, T_p and B_t are constants, $T_{ca} + T_p = 1.25ms$ and $B_t = 8224$ bits [Erc08].

The airtime cost of an AP a is the aggregated airtime cost of its associated STAs, that is,

$$T(a, t) = \sum_{s \in S_a} T_s^a(t) = \sum_{s \in \mathscr{S}_i} \left[T_{ca} + T_p + \frac{B_t}{C_s^a(t)} \right], \tag{8.2}$$

which is a good approximation of the uplink transmission delay due to the contention among STAs associated with the same AP [Erc08].

In high-density WiFi networks, the number of orthogonal channels can be far less than the number of neighboring APs. As a result, multiple neighboring APs may have to share the same channel, the contention is not only from the STAs of the same AP, but also from other co-channel neighboring APs and their associated STAs. Therefore, the airtime cost should take into account the co-channel contention delay from neighboring APs. Specifically, for an AP a operating in channel i, its airtime cost is given by:

$$T(a, i, t) = T(a, t) + \sum_{a' \in \mathscr{N}_a^i} T(a', t)$$

$$= \sum_{a' \in \mathscr{N}_a^i \bigcup \{a\}} \sum_{s \in \mathscr{S}_j} [T_{ca} + T_p + \frac{B_t}{C_s^{a'}(t)}], \tag{8.3}$$

where $\mathscr{N}_a^i$ denotes the set of neighboring APs operating in the same channel as AP a.

8.2.3 Channel Selection and User Association (CSUA) Problem

As seen from (8.3), the airtime cost is dominated by two factors, one is the intra-AP contention between STAs associated with the same AP; the other is the inter-AP contention between co-channel neighboring APs. Therefore, from an AP's perspective, an effective way to mitigate co-channel contention is to select a less congested channel so that its airtime cost is minimized, while from a STA's perspective, it prefers an AP with smaller airtime cost so that the overall contention is minimized. Mathematically, we have the following channel selection problem for an AP a:

$$i_a^* = \arg \min_{i=1,\cdots,M} T(a, i, t), \tag{8.4}$$

and the user association problem for a STA s:

$$a_s^* = \arg \min_{a \in \mathscr{A}_s} T(a, i, t). \tag{8.5}$$

Unfortunately, the airtime cost is unknown a priori. The reason is twofold. First, the transmission rate between a STA s and its associated AP a depends on the its perceived signal-to-noise ratio (SNR), i.e., $C_s^a(t) = f(SNR_s^a)$. For example, for AWGN channels, the transmit rate can be upper bounded by the Shannon's equation, that is, $C_s^a(t) \leq B \log \left(1 + \frac{P_s G_s^a(t)}{N_0}\right)$, where B is the channel bandwidth in hertz, P_s is the transmit power, $G_s^a(t)$ is the instantaneous channel gain at time t, and N_0 is the noise power. Due to fast-fading and shadowing effects, the channel gain changes over time following an unknown distribution, and thus $C_s^a(1), \cdots, C_s^a(t)$ can be modeled as IID random variables. Second, the co-channel contention depends on the number of neighboring APs and STAs operating in the same channel, which is unknown since the decisions of other APs and STAs are made online.

Therefore, the CSUA problem is nontrivial since the airtime cost *can only be estimated after all APs and STAs has made their decision*. In next section, we adopt the adversarial MAB framework to address this problem.

8.3 Adversarial MAB Framework for CSUA Problem and Algorithms

In this section, we first introduce the adversarial MAB framework for the CSUA problem, and then propose an algorithm based on the exponentially weighted average strategy.

8.3.1 Adversarial MAB Formulation

The CSUA problem can be modeled as an adversarial MAB problem consisting of two kinds of players (APs and STAs), who can choose an action from a set of actions (a set of M channels for an AP a, and a set of APs $\mathscr{A}_s$ for a STA s). This problem can be seen as a game with two kinds of agents: for an AP a (or a STA s), the first agent is itself, the second agent is the set of all other players whose actions (channel selection or user association) affect its airtime cost.

In the following, we focus on the channel selection problem from an AP's perspective. The same strategy is applicable to the user association problem as well. Specifically, each AP a chooses a channel i in successive rounds. Upon operating on channel i in round t, AP a observes its airtime cost $T(a, i, t)$, while no other information can be accessed. In order to measure the utility of a channel selection sequence, the concept of *regret* is adopted to denote the difference between the airtime cost which could have been achieved if the AP selects the optimal channel, and the airtime cost which is achieved in the actually selected channel. The objective of AP a is to minimize its cumulative regret $R_a(n)$ up to time n, which is defined as follows:

$$R_a(n) = \max_{i=1,\cdots,M} \sum_{t=1}^{n} W(a, i, t) - \sum_{t=1}^{n} W(a, I_t^{(a)}, t), \tag{8.6}$$

where $I_t^{(a)}$ denotes the channel selected by AP a at time t and $W(a, i, t) = \frac{1}{T(a,i,t)}$ denotes the reward of channel i selected by AP a at time t.

This definition can be generalized to the case where AP a selects channel using a mixed strategy, which is characterized by a probability distribution $\mathbf{P_t} = (p_{1,t}^{(a)}, p_{2,t}^{(a)}, \ldots, p_{M,t}^{(a)})$ over the set of M possible channels, where each channel i is selected with probability $p_{i,t}^{(a)}$. In this case, the concept of *external regret* can be adopted to measure the difference between the expected cost of the actual mixed strategy and the optimal channel selection:

$$\begin{aligned}
R_a^{Ext}(n) &= \max_{i=1,\cdots,M} \sum_{t=1}^{n} W(a, i, t) - \sum_{t=1}^{n} \bar{W}(a, \mathbf{P_t}, t) \\
&= \max_{i=1,\cdots,M} \sum_{t=1}^{n} \sum_{j=1}^{M} p_{j,t}^{(a)} (W(a, i, t) - W_a(a, j, t))
\end{aligned} \tag{8.7}$$

where $\bar{W}(a, \mathbf{P_t}, t) = E_t[W(a, \mathbf{P_t}, t)] = \sum_{j=1}^{M} [p_{j,t}^{(a)} W(a, j, t)]$ denotes the expected airtime cost for AP a to operate on a channel according to the probability distribution $\mathbf{P_t}$.

In order to compare the channel selection in pair, we adopt *internal regret* to measure the expected regret by choosing channel i instead of channel j, which is given by:

$$\begin{aligned}
R_a^{Int}(n) &= \max_{i,j=1,\cdots,M} r_{(i \to j),n}^{(a)} \\
&= \max_{i,j=1,\cdots,M} \sum_{t=1}^{n} p_{i,t}^{(a)} [W(a, j, t) - W(a, i, t)]
\end{aligned} \tag{8.8}$$

The relationship between internal regret and Nash equilibrium is established in Theorem 4.10.

8.3.2 Algorithm

Algorithm 8.1: EWA_CS algorithm

---

1 **Init:** *Set* $\gamma_t = t^{-\frac{1}{3}}$ *and* $\eta_t = \frac{\gamma_t^3}{M^2}$;

2 Construct a modified strategy for each pair of channels i and j as follows:

$P^{(a)}_{i \to j, t-1} = (p^{(a)}_{1,t-1}, \cdots, 0, \cdots, p^{(a)}_{j,t-1} + p^{(a)}_{i,t-1}, \cdots, p^{(a)}_{M,t-1})$, where $p^{(a)}_{i,t-1}$ is replaced by 0, and $p^{(a)}_{j,t-1}$ is replaced by $p^{(a)}_{j,t-1} + p^{(a)}_{i,t-1}$;

3 Update

$$\Delta^{(a)}_{i \to j,t} = \frac{\exp(\eta_t \tilde{r}^{(a)}_{i \to j,t-1})}{\sum_{k \to l, k \neq l} \exp(\eta_t \tilde{r}^{(a)}_{k \to l,t-1})} \tag{8.9}$$

4 where

$$\tilde{r}^{(a)}_{i \to j,t-1} = \sum_{\tau=1}^{t-1} p^{(a)}_{i,\tau} [\tilde{W}(a,j,\tau) - \tilde{W}(a,i,\tau)] \tag{8.10}$$

5 and for $k = 1, \cdots, M$:

$$\tilde{W}(a,k,\tau) = \begin{cases} \frac{W(a,k,\tau)}{p^{(a)}_{k,\tau}}, & \text{if } k = I^{(a)}_\tau. \\ 0, & \text{otherwise.} \end{cases} \tag{8.11}$$

6 Solve the following fixed point equation to find $P^{(a)}_t$

$$P^{(a)}_t = \sum_{(i,j):i \neq j} P^{(a)}_{i \to j,t} \Delta^{(a)}_{i \to j,t} \tag{8.12}$$

7 Finally yields:

$$P^{(a)}_t = (1 - \gamma_t) P^{(a)}_t + \frac{\gamma_t}{M}. \tag{8.13}$$

8 Select a channel $I^{(a)}_t \in \{1, \cdots, M\}$ according to $P^{(a)}_t$;

We propose algorithms for the CSUA problem based on the bandit version of the internal regret minimizing algorithm with exponentially weighted average strategy as introduced in Sect. 4.2.4. The core idea of the algorithm is that each player chooses an action using a mixed strategy where the probabilities are adjusted according to the internal regret. In order to guarantee that the correlated equilibria can be obtained, the algorithm needs to achieve vanishing internal regret as discussed in Sect. 4.3. To this end, the algorithm needs to maintain two parameters, γ_t and η_t, which guarantees the Hannan consistency and leads to vanishing internal regrets [CL06].

Algorithm 8.2: CSUA Algorithm

---

1 **for** $t = 1, \cdots$ **do**
2   **for** $a \in \mathscr{A}$ **do**
3   ⌊ Call EWA_CS algorithm to select a channel for AP a;
4   **for** $s \in \mathscr{S}$ **do**
5   ⌊ Call EWA_UA algorithm to select an AP to associate for STA s;
6   **for** *all players* **do**
7   ⌊ Play the selected actions and observe the airtime costs;

The algorithm consists of two stages, that is, a channel selection stage and a user association stage. In each round t, the channel selection stage is executed first, whereby each AP a runs an EWA based channel selection algorithm (EWA_CS) to select a channel and observes the corresponding airtime cost. The details of the EWA_CS algorithm is shown in Algorithm 8.1. First, the parameters are set as $\gamma_t = t^{-\frac{1}{3}}$ and $\eta_t = \frac{\gamma_t^3}{M^2}$, and a modified strategy is constructed based on the probability distribution $\mathbf{P}_{t-1}^{(a)} = (p_{1,t-1}^{(a)}, p_{2,t-1}^{(a)}, \ldots, p_{M,t-1}^{(a)})$, which yields a probability $P_{i \to j,t-1}^{(a)}$ for each pair of i and j by transferring the probability mass from i to j (line 2), with which the internal regret $\tilde{r}_{i \to j,t}^{(a)}$ and a weighted parameter $\Delta_{i \to j,t}^{(a)}$ can be obtained (line 3–4). Note that since each AP can only observe the airtime cost of the selected channel but not those of others, the airtime cost can only be estimated using an unbiased estimator (8.11) (line 5), that is $E_t(\widetilde{W}(a, k, t)) = W(a, k, t)$, where E denotes the expectation operator. The probability distribution $P_t^{(a)}$ is obtained by solving a set of fixed equations (line 6) and finally adjusted using the EWA strategy (line 7).

After all APs have selected their channels, the algorithm enters the user association stage, where each STA s runs an EWA-based user association algorithm (EWA_UA) to select its AP. The procedure is similar to the EWA_CS algorithm as shown in Algorithm 8.1. The action set for each STA s is its neighboring AP set $\mathscr{A}_s$, and the parameters are set as $\gamma_t = t^{-\frac{1}{3}}$ and $\eta_t = \frac{\gamma_t^3}{|A_s|^2}$. Once all players have made their decisions, the airtime costs are observed. The algorithm enters the next round and the same procedure is repeated. The overall procedure of the CSUA algorithm is shown in Algorithm 8.2.

Using a similar technique for Theorem 6.9 of [CL06], it can be proven that this algorithm have *vanishing internal regrets*, and the joint distribution of plays converges to a set of correlated equilibria.

Table 8.1 SINR versus link capacity in IEEE 802.11 a/g

Rate Index	0	1	2	3	4	5	6	7
SINR (dB)	6	7.8	9	10.8	17	18.8	24	24.6
Rate (Mbps)	6	9	12	18	24	36	48	54

8.4 Performance Evaluation

8.4.1 Simulation Setup

In this section, we study the convergence of the CSUA algorithm and its performance under different settings. The network considered in the simulations consists of a set of APs and STAs, distributed randomly in 1000 m × 1000 m rectangular area. The large scale path loss model is adopted with the path loss exponent set to 4. The transmit power of each AP is set to 15 dBm, and the noise level is assumed to be constant and set to −95 dBm. The transmit rate between a STA and an AP is determined by the SNR of the link according to the IEEE 802.11a/g protocol as shown in Table 8.1. The maximum transmit range is 300 m, and the contention range is 500 m.

8.4.2 Simulation Results

First, we study the convergence of the CSUA algorithm. In Fig. 8.2, we show the evolution of the channel selection probability vector for a specific AP. It can be seen that the channel selection probability vector converges to (1,0,0) in about 600 simulation rounds. In other words, the first channel is determined by the channel selection algorithm, which is in fact the optimal solution in equilibrium. Figure 8.3 shows the convergence of the user association probability vector for a specific STA, which has 4 APs in its transmit range. It can be seen that the probability vector converges to (0, 1, 0, 0) eventually, which suggests that the proposed user association algorithm is effective in finding the optimal AP in equilibrium. In Fig. 8.4, we shows the convergence of the airtime cost of three APs. It can be seen that the airtime cost varies dramatically at the beginning. However, as the channel selection of each AP converges, the airtime cost also becomes stable.

Second, we study the performance of the CSUA algorithm under different settings. In Figs. 8.5, 8.6 and 8.7, we show the airtime cost of APs under different number of APs, STAs and channels respectively, where the airtime cost is obtained after the simulation is ran for 1500 rounds. In these figures, the horizontal lines are the average airtime of all APs, and the error bar indicates the maximum and minimum airtime costs of all APs.

Fig. 8.2 Convergence of the
channel selection probability
vector

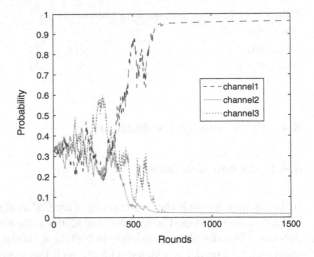

Fig. 8.3 Convergence of the
user association probability
vector

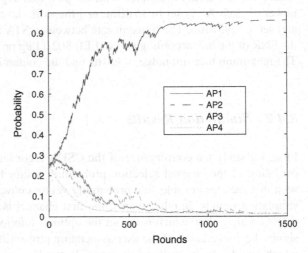

In Fig. 8.5, we show the airtime costs with 125 STAs and 3 channels, while the
number of APs varies from 5 to 25. It can be seen that the airtime cost is gradually
decreasing with the increase of APs. The reason is twofold. First, as more APs are
deployed in the network area, the traffic loads of existing APs are alleviated. Second,
the transmit rate increases as the STAs and APs are closer, which in turn reduces the
airtime cost. However, when the number of APs is beyond 15, the performance gain
is compromised due to the increasing contention among APs at a fixed transmission
power level. Therefore, the airtime cost cannot be further improved.

In Fig. 8.6, we show the airtime cost with 10 APs and 3 channels, and the number
of STAs varies from 50 to 200. It can be seen that the airtime cost increases gradually
since the average traffic load of APs is increased as the user density increases.

Fig. 8.4 Convergence of the airtime cost

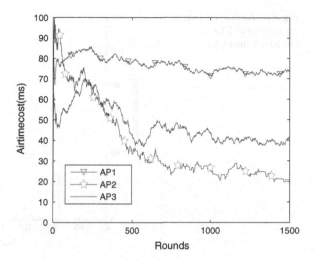

Fig. 8.5 Airtime cost versus the number of APs (125 STAs, 3 channels)

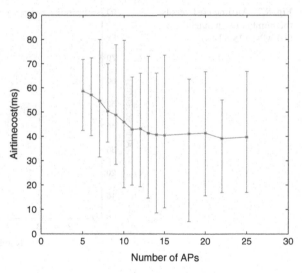

Finally, Fig. 8.7 shows the airtime cost with 10 APs and 125 STAs, and the number of channel varies from 2 to 8. It can be seen that the increase of channels can help mitigate co-channel interference, which leads to the decrease of the airtime cost. On the other hand, as the number of channels is beyond 5, the performance gain is no longer significant, which suggests that the contention among APs can be well controlled by the channel allocation strategy given sufficient number of channels.

Fig. 8.6 Airtime cost versus
the number of STAs
(10 APs,3 channels)

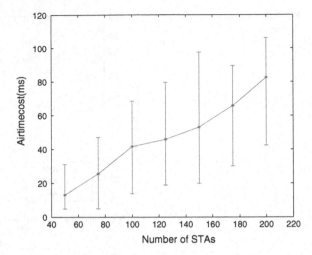

Fig. 8.7 Airtime cost versus
the number of channels
(10 APs, 125 STAs)

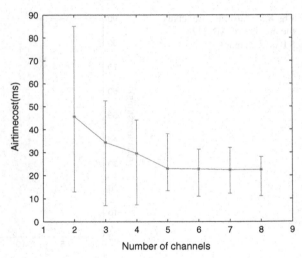

8.5   Summary

In this chapter, we adopted the adversarial MAB framework to address the channel
selection and user association problem in high-density WiFi networks, which captured both the uncertainty of channel states and the selfishness of individual STAs
and APs. An exponentially weighted average strategy was adopted to design the
algorithms for this problem, which is guaranteed to converge to a set of correlated
equilibria with vanishing regrets. Simulation results also showed the performance of
the proposed scheme under different network settings.

References

[C11]   Cisco Connected Stadium Wi-Fi for Sports and Entertainment Venues. Tech.
            rep. http://www.cisco.com/c/dam/en/us/products/collateral/wireless/aironet-3500p-
            access-point/white_paper_c11-674354.pdf. 2011.
[BHL04]   Yigal Bejerano, Seung-Jae Han, and Li Erran Li. "Fairness and load balancing in wire-
            less LANs using association control". In: Proceedings of the 10th annual international
            conference on Mobile computing and networking. ACM. 2004, pp. 315–329.
[CHD10]   Surachai Chieochan, Ekram Hossain, and Jeffrey Diamond. "Channel assignment
            schemes for infrastructure-based 802.11 WLANs: A survey". In: Communications
            Surveys & Tutorials, IEEE 12.1 (2010), pp. 124–136.
[CL06]   N. Cesa-Bianchi and G. Lugosi. "Prediction, Learning, and Games". In: Cambridge
            University Press. 2006.
[Erc08]   Ozgur Ercetin. "Association games in IEEE 802.11 wireless local area networks". In:
            Wireless Communications, IEEE Transactions on 7.12 (2008), pp. 5136–5143.
[Hai+08]   Mohamad Haidar et al. "Channel Assignment in an IEEE 802.11 WLAN based on
            Signal-to-Interference Ratio". In: Electrical and Computer Engineering, 2008. CCECE
            2008. Canadian Conference on. IEEE. 2008, pp. 001169–001174.
[IEE08]   IEEE. IEEE 802.11s draft standard: ESS extended service set mesh networking. Tech.
            rep. 2008.
[IEE14]   IEEE. IEEE 802.11 High Efficiency WLAN (HEW) Study Group. Tech. rep. 2014.
[Kau+07]   B. Kauffmann et al. "Measurement-Based Self Organization of Interfering 802.11 Wire-
            less Access Networks". In: INFOCOM 2007. May 2007, pp. 1451–1459.
[XHH11]   Wenchao Xu, Cunqing Hua, and Aiping Huang. "Channel assignment and user asso-
            ciation game in dense 802.11 wireless networks". In: Communications (ICC), 2011
            IEEE International Conference on. IEEE. 2011, pp. 1–5.
[YMS08]   Ming Yu, Aniket Malvankar, and Wei Su. "A distributed radio channel allocation scheme
            for WLANs with multiple data rates". In: Communications, IEEE Transactions on 56.3
            (2008), pp. 454–465.
[Yu+07]   Ming Yu et al. "A new radio channel allocation strategy for WLAN APs with power con-
            trol capabilities". In: Global Telecommunications Conference, 2007. GLOBECOM'07.
            IEEE. IEEE. 2007, pp. 4893–4898.
[Yua+13]   Wei Yuan et al. "Variable-width channel allocation for access points: A game-theoretic
            perspective". In: Mobile Computing, IEEE Transactions on 12.7 (2013), pp. 1428–
            1442.

Index

© Springer International Publishing AG 2016
R. Zheng and C. Hua, *Sequential Learning and Decision-Making in Wireless Resource Management*, Wireless Networks, DOI 10.1007/978-3-319-50502-2

Printed in the United States
By Bookmasters

Printed in the United States
By Bookmasters